MAXIMUM PRINCIPLES FOR THE HILL'S EQUATION

MAXIMUM PRINCIPLES FOR THE HILL'S EQUATION

Alberto Cabada
Universidade de Santiago de Compostela,
Instituto de Matemáticas,
Facultade de Matemáticas,
Santiago de Compostela, Galicia, Spain

José Ángel Cid
Universidade de Vigo,
Departamento de Matemáticas,
Ourense, Galicia, Spain

Lucía López-Somoza
Universidade de Santiago de Compostela,
Instituto de Matemáticas,
Facultade de Matemáticas,
Santiago de Compostela, Galicia, Spain

Academic Press is an imprint of Elsevier
125 London Wall, London EC2Y 5AS, United Kingdom
525 B Street, Suite 1800, San Diego, CA 92101-4495, United States
50 Hampshire Street, 5th Floor, Cambridge, MA 02139, United States
The Boulevard, Langford Lane, Kidlington, Oxford OX5 1GB, United Kingdom

Library of Congress Cataloging-in-Publication Data
A catalog record for this book is available from the Library of Congress

British Library Cataloguing-in-Publication Data
A catalogue record for this book is available from the British Library

ISBN: 978-0-12-804117-8

For information on all Academic Press publications
visit our website at https://www.elsevier.com/books-and-journals

Working together
to grow libraries in
developing countries

www.elsevier.com • www.bookaid.org

Publisher: Candice Janco
Acquisition Editor: Graham Nisbet
Editorial Project Manager: Susan Ikeda
Production Project Manager: Omer Mukthar
Designer: Mark Rogers

Typeset by VTeX

This book is dedicated to my brother and sisters José Luis, Marina, Mercedes, Nieves and Victoria.

Alberto Cabada

This book is dedicated to my wife Natalia and our children, Gael and Noa.

José Ángel Cid

This book is dedicated to my parents, Deme and Aurora, and my brother Pablo.

Lucía López-Somoza

CONTENTS

About the Authors		*ix*
Preface		*xi*
Acknowledgment		*xiii*

1. Introduction
1.1.	Hill's Equation	1
1.2.	Stability in the Sense of Lyapunov	4
1.3.	Floquet's Theorem for the Hill's Equation	8
	References	18

1. Introduction — **1**

2. Homogeneous Equation — 19
2.1.	Introduction	19
2.2.	Sturm Comparison Theory	20
2.3.	Spectral Properties of Dirichlet Problem	26
2.4.	Spectral Properties of Mixed and Neumann Problems	29
2.5.	Spectral Properties of the Periodic Problem: Intervals of Stability and Instability	31
2.6.	Relation Between Eigenvalues of Neumann, Dirichlet, Periodic, and Antiperiodic Problems	42
	References	47

3. Nonhomogeneous Equation — 49
3.1.	Introduction	49
3.2.	The Green's Function	52
3.3.	Periodic Conditions	58
3.3.1.	Properties of the Periodic Green's Function	58
3.3.2.	Optimal Conditions for the Periodic MP and AMP	63
3.3.3.	Explicit Criteria for the Periodic AMP and MP	68
3.3.4.	More on Explicit Criteria	72
3.3.5.	Examples	75
3.4.	Non-Periodic Conditions	80
3.4.1.	Neumann Problem	81
3.4.2.	Dirichlet Problem	84
3.4.3.	Relation Between Neumann and Dirichlet Problems	85
3.4.4.	Mixed Problems and their Relation with Neumann and Dirichlet Ones	86
3.4.5.	Order of Eigenvalues and Constant Sign of the Green's Function	91
3.4.6.	Relations Between Green's Functions. Comparison Principles	100
3.4.7.	Constant Sign for Non-Periodic Green's Functions	103

	3.4.8.	Global Order of Eigenvalues	107
	3.4.9.	Examples	109
3.5.	General Second Order Equation		113
	3.5.1.	Periodic Problem	113
	3.5.2.	Non-Periodic Conditions	120
	References		127

4.	**Nonlinear Equations**		**129**
	4.1.	Introduction	129
	4.2.	Fixed Point Theorems and Degree Theory	130
		4.2.1. Leray-Schauder Degree	131
		4.2.2. Fixed Point Theorems	134
		4.2.3. Extremal Fixed Points	148
		4.2.4. Monotone Operators	150
		4.2.5. Non-increasing Operators	159
		4.2.6. Non-decreasing Operators	163
		4.2.7. Problems with Parametric Dependence	168
	4.3.	Lower and Upper Solutions Method	184
		4.3.1. Well Ordered Lower and Upper Solutions	187
		4.3.2. Existence of Extremal Solutions	198
		4.3.3. Non-Well-Ordered Lower and Upper Solutions	201
	4.4.	Monotone Iterative Techniques	209
		4.4.1. Well Ordered Lower and Upper Solutions	210
		4.4.2. Reversed Ordered Lower and Upper Solutions	213
	References		218

A.	**Sobolev Inequalities**		**223**
	References		232

Glossary		*235*
Index		*237*

ABOUT THE AUTHORS

Alberto Cabada is Professor at the University of Santiago de Compostela. His line of research is devoted to nonlinear differential equations. He has obtained some results of existence and multiplicity of solution in differential equations, both ordinary and partial, as well as difference equations and fractional ones. The techniques used are mainly based on topological methods and iterative techniques. An important part of his research focuses on the study of, both quantitative and qualitative, properties of the so-called Green's functions. He is the author of more than one hundred thirty research articles indexed in the Citation Index Report and has authored two monographs. He has supervised several Master and Ph.D. students and has been the lead of different academic institutions as the Department of Mathematical Analysis and the Institute of Mathematics of the University of Santiago de Compostela.

José Ángel Cid is Associate Professor at the University of Vigo. His research is focused in the field of ordinary differential equations. He has dealt mainly with qualitative properties like existence, uniqueness and multiplicity, obtained by means of topological and variational methods, fixed point theory and monotone iterative techniques. He has authored more than forty research papers included in the Citation Index Report. He has taught at the universities of Santiago de Compostela, Jaén and Vigo.

Lucía López-Somoza is a Ph.D. student at University of Santiago de Compostela. Her line of research is the study of nonlinear functional differential equations. She studies Hill's equation and, in particular, the relations between the solutions of this equation under different types of boundary conditions. At present she is a research fellowship at University of Santiago de Compostela.

PREFACE

This book is devoted to the study of basic properties of the Hill's equation, both in homogeneous and non homogeneous cases. As regards the homogeneous problem, the spectral problem will be treated, along with the oscillation of the solutions and their stability. Concerning the non homogeneous problem, we will consider comparison principles for the Hill's equation. More concisely it will be delivered to the properties of the Green's functions related to such equation coupled with different boundary value conditions. We will establish its relationship with the spectral theory developed for the homogeneous case. So stability and constant sign solutions of the equation will be considered. Classical and recent results obtained by us and another authors will be presented.

The existence of solutions of nonlinear boundary value problems will be also studied. The used techniques will mainly consist on the construction of integral operators defined on abstract spaces whose fixed points coincide with the solutions of the nonlinear problems that we are considering. So, the fundamental construction of the classical Leray–Schauder degree will be shown, and classical fixed point theorems will be deduced. We will make special emphasis on operators defined on cones which, as we will see, allow us to find constant sign solutions. Moreover, the theory of lower and upper solutions and the monotone iterative techniques will be also developed in a general framework and applied to nonlinear problems related to the Hill's equation.

This book is directed to a wide range of mathematicians, including both theoretical and applied oriented ones, working on the subject of differential equations. The book also could be used for a Ph. D course addressed to graduate students.

The audience will benefit of a short book providing both complete and accessible information of classical results and recent developments related to the subject.

Alberto Cabada, José Ángel Cid, Lucía López-Somoza
Ourense and Santiago de Compostela

September 2017

ACKNOWLEDGMENT

We thank the editorial team at Elsevier, specially Mr. Graham Nisbet, Senior Acquisitions Editor, and Ms. Susan Ikeda, Editorial Project Manager, for guidance throughout the publishing process.

We also thank to Pr. F. Adrián F. Tojo for his interesting suggestions in the preparation of this manuscript.

This book was supported by Ministerio de Economía y Competitividad, Spain, and FEDER, project MTM2013-43014-P, Agencia Estatal de Investigación (AEI) of Spain under grant MTM2016-75140-P, co-financed by the European Community fund FEDER, and by Xunta de Galicia (Spain), project EM2014/032.

Alberto Cabada, José Ángel Cid, Lucía López-Somoza

Ourense and Santiago de Compostela

September 2017

CHAPTER 1

Introduction

Contents

1.1 Hill's Equation   1
1.2 Stability in the Sense of Lyapunov   4
1.3 Floquet's Theorem for the Hill's Equation   8
References   18

1.1 HILL'S EQUATION

The Hill's equation,

$$u''(t) + a(t)\, u(t) = 0, \tag{1.1}$$

has numerous applications in engineering and physics. Among them we can find some problems in mechanics, astronomy, circuits, electric conductivity of metals and cyclotrons. Hill's equation is named after the pioneering work of the mathematical astronomer George William Hill (1838–1914), see [6]. He also made contributions to the three and the four body problems.

Moreover, the theory related to the Hill's equation can be extended to every differential equation in the form

$$u''(t) + p(t)\, u'(t) + q(t)\, u(t) = 0, \tag{1.2}$$

such that the coefficients p and q have enough regularity. This is due to the fact that, with a suitable change of variable, the previous equation transforms in one of the type of (1.1) (see the details in Section 2.2 of Chapter 2).

As a first example we could consider a mass–spring system, that is, a spring with a mass m hanging from it. It is very well-known that, denoting by $u(t)$ the position of the mass at the instant t and assuming absence of friction, the previous model can be expressed as

$$u''(t) + \frac{k}{m}\, u(t) = 0,$$

with $k > 0$ the elastic constant of the spring.

However, in a real physical system, there exists a friction force which opposes the movement and is proportional to the object's speed. In this case

Maximum Principles for the Hill's Equation.
DOI: http://dx.doi.org/10.1016/B978-0-12-804117-8.00001-1
Copyright © 2018 Elsevier Inc. All rights reserved.

the situation can be modelled by the equation

$$u''(t) + \mu\, u'(t) + \frac{k}{m}\, u(t) = 0,$$

with μ the so-called friction coefficient. The value of such coefficient is characteristic of the environment where the object oscillates, and depends, among other variables, on the density, temperature, and pressure of the environment. However, it could be considered a situation in which the spring moves between two different environments, each one with its particular friction coefficient. Also, the environment could have strong variations of density or temperature that could cause changes in the friction coefficient depending on time. This could be modelled by substituting the friction coefficient μ for a not necessarily constant function $\mu(t)$

$$u''(t) + \mu(t)\, u'(t) + \frac{k}{m}\, u(t) = 0.$$

Another possible situation would be that one in which there exists another external force acting periodically on the mass in such a way that it tends to move the mass back into its position of equilibrium, acting in proportion to the distance to that position. Including this new term in the previous model we have

$$u''(t) + \mu(t)\, u'(t) + \left(\frac{k}{m} + F(t) \right) u(t) = 0.$$

In any of the two cases, we obtain an equation in the form (1.2) in which, if $\mu(t)$ has enough regularity, we could do the following change of variable

$$a(t) = \frac{k}{m} + F(t) - \frac{1}{4} \left(\frac{\mu(t)}{m} \right)^2 - \frac{1}{2} \frac{\mu'(t)}{m}$$

and transform the equation in one in the form (1.1).

A second example studied in [4,8] is the inverted pendulum. A mathematical pendulum consists of a particle of mass m connected to a base through a string (which is supposed to be rigid and of despicable weight) in such a way that the mass moves in a fixed vertical plane. If the particle moves by the force of gravity, then the movement of the pendulum is given by the equation

$$\theta''(t) - \frac{g}{l} \sin\left(\theta(t) \right) = 0,$$

where g denotes the gravity, l the length of the string, and θ represents the angle between the string and the perpendicular line to the base.

In the surroundings of the equilibrium point $\theta = 0$, we can approximate $\sin(\theta) \approx \theta$ so the equation of movement could be rewritten as

$$\theta''(t) - \frac{g}{l}\theta(t) = 0.$$

Consider now the case in which the suspension point of the string vibrates vertically with an acceleration $a(t)$. Then, as it is proved in [4], the equation of movement would change into

$$\theta''(t) - \frac{1}{l}(g + a(t))\theta(t) = 0,$$

which is of the form (1.1).

Other equations which fit into the framework of the Hill's equation are the following ones:

• Airy's equation: (see [13])

$$u''(t) + t\,u(t) = 0.$$

This equation appears in the study of the diffraction of light, the diffraction of radio waves around the Earth's surface, in aerodynamics and in the swing of a uniform vertical column which bounds under its own weight.

• Mathieu's equation: (see [2,14,15])

$$u''(t) + (e + b\cos t)\,u(t) = 0.$$

It is the result of the analysis of the phenomenon of parametric resonance associated with an oscillator whose parameters change with time. It appears in problems related to periodic movements, as the trajectory of an electron in a periodic arrange of atoms.

When studying oscillation phenomena of the solutions of (1.1), it is observed that these are determined by the potential a. In particular, solutions of (1.1) do not oscillate when $a < 0$ but they do it infinitely many times for $a > 0$ large enough. Moreover, the larger the potential a is, the faster the solutions of (1.1) oscillate.

By simply considering that every integrable function can be rewritten as

$$a(t) = \bar{a} + \hat{a}(t),$$

where $\bar{a} = \dfrac{1}{T}\displaystyle\int_0^T a(s)\,ds$ and $\hat{a}(t) = a(t) - \bar{a}$, it is obvious that studying the potentials a for which the solutions of Eq. (1.1) oscillate in $[0, T]$, is equivalent to study the values of $\lambda \in \mathbb{R}$ for which the equation

$$u''(t) + (a(t) + \lambda)\,u(t) = 0, \quad t \in [0, T], \tag{1.3}$$

with $a \in L^\alpha[0, T]$ fixed, $\alpha \geq 1$, has no trivial solution.

If we consider Eq. (1.3) submitted to boundary value conditions, we have a spectral problem.

First studies about the Hill's equation were focused on the homogeneous case, from the point of view of the classical oscillation theory of Sturm-Liouville ([5,9,13]). In particular, from the study of Eq. (1.3) under periodic boundary conditions, important results related to the stability of solutions were obtained.

Afterwards the nonhomogeneous periodic problem,

$$u''(t) + (a(t) + \lambda)\,u(t) = \sigma(t), \; t \in [0, T], \quad u(0) = u(T), \quad u'(0) = u'(T),$$

was studied, with $a \in L^\alpha[0, T]$, $\alpha \geq 1$, and $\sigma \in L^1[0, T]$. In this case it results specially interesting the study of constant sign solutions when σ does not change sign. This situation could be interpreted in a physical way by considering σ as an external force acting over the system; then constant sign solutions would mean that a positive perturbation maintains oscillations above or below the equilibrium point. The study of constant sign solutions carried to consider comparison principles (MP and AMP) which, later, were related to the constant sign of the Green's function.

1.2 STABILITY IN THE SENSE OF LYAPUNOV

In this section we will restrict ourselves to the concept of stability in the case of the Hill's equation (1.1), although this concept can be stated for the solution of any differential equation, see [3]. Firstly, consider the potential a in (1.1) extended periodically to $\mathbb{R}$. Since (1.1) is linear it is well-known that for each initial conditions $u(t_0) = u_0$ and $u'(t_0) = \bar{u}_0$ there exists a unique solution which is defined for all $t \in \mathbb{R}$. We will denote the unique solution of

$$u''(t) + a(t)\,u(t) = 0, \; t \in \mathbb{R}, \; u(t_0) = u_0, \; u'(t_0) = \bar{u}_0, \tag{1.4}$$

by $u(t, t_0, u_0, \bar{u}_0)$ in order to stress its dependence on the initial conditions.

Definition 1 (Stability in the sense of Lyapunov). The solution $u(\cdot, t_0, u_0, \bar{u}_0)$ of (1.4) is stable if and only if for every $\varepsilon > 0$ there exists $\delta > 0$ such that if

$$\max\{|u_1 - u_0|, |\bar{u}_1 - \bar{u}_0|\} \leq \delta,$$

then

$$\max\{|u(t, t_0, u_1, \bar{u}_1) - u(t, t_0, u_0, \bar{u}_0)|, |u'(t, t_0, u_1, \bar{u}_1) - u'(t, t_0, u_0, \bar{u}_0)|\} \leq \varepsilon$$

for all $t \in \mathbb{R}$.

Definition 2. The solution $u(\cdot, t_0, u_0, \bar{u}_0)$ of (1.4) is bounded (or C^1 bounded) if and only if there exists a constant $M > 0$ such that

$$\max\{|u(t, t_0, u_0, \bar{u}_0)|, |u'(t, t_0, u_0, \bar{u}_0)|\} \leq M$$

for all $t \in \mathbb{R}$.

In general the concepts of stability and boundedness for differential equations are different, as the simple equation $u' = 1$ shows: all its solutions $u(t, t_0, u_0) = t + (u_0 - t_0)$ are stable but unbounded. However, in the linear case, and in particular for problem (1.4), we have the following useful result.

Theorem 1. *The following claims are equivalent:*
1. *The solution $u(\cdot, t_0, u_0, \bar{u}_0)$ of (1.4) is stable for each $(u_0, \bar{u}_0) \in \mathbb{R}^2$.*
2. *The solution $u(\cdot, t_0, 0, 0) \equiv 0$ of (1.4) is stable.*
3. *The solution $u(\cdot, t_0, u_0, \bar{u}_0)$ of (1.4) is bounded for each $(u_0, \bar{u}_0) \in \mathbb{R}^2$.*

Proof. Firstly, we are going to prove that 1 and 2 are equivalent. Taking into account that, by linearity,

$$u(\cdot, t_0, u_1 + u_0, \bar{u}_1 + \bar{u}_0) = u(\cdot, t_0, u_1, \bar{u}_1) + u(\cdot, t_0, u_0, \bar{u}_0),$$

we have that

$$\max\{|u(t, t_0, u_1 + u_0, \bar{u}_1 + \bar{u}_0) - u(t, t_0, u_0, \bar{u}_0)|,$$
$$|u'(t, t_0, u_1 + u_0, \bar{u}_1 + \bar{u}_0) - u'(t, t_0, u_0, \bar{u}_0)|\}$$
$$= \max\{|u(t, t_0, u_1, \bar{u}_1)|, |u'(t, t_0, u_1, \bar{u}_1)|\}.$$

Thus it is clear that the stability at any point $(u_0, \bar{u}_0) \in \mathbb{R}^2$ is equivalent to the stability at $(0, 0)$.

Now, we will show that 2 and 3 are equivalent. Again by linearity we have

$$u(\cdot, t_0, u_0, \bar{u}_0) = u_0\, u(\cdot, t_0, 1, 0) + \bar{u}_0\, u(\cdot, t_0, 0, 1).$$

Then the stability at $(0,0)$ is clearly equivalent to the boundedness of the particular solutions $u(\cdot, t_0, 1, 0)$ and $u(\cdot, t_0, 0, 1)$ which, again by linearity, is equivalent to the boundedness of the solution at any point $(u_0, \bar{u}_0) \in \mathbb{R}^2$.

$\square$

Definition 3. In view of Theorem 1, we will say that the trivial solution of (1.1) is stable, or simply that Eq. (1.1) is stable, if and only if all the solutions of (1.1) are bounded in C^1.

Otherwise, we will say that Eq. (1.1), or the trivial solution of (1.1), is unstable.

Floquet's theory for first order linear differential systems with T-periodic coefficients, see [3,5,12], tells us that the stability of Eq. (1.1) is determined by the eigenvalues of the matrix $M(T)$, where

$$M(t) := \begin{pmatrix} u_1(t) & u_2(t) \\ u_1'(t) & u_2'(t) \end{pmatrix}, \tag{1.5}$$

and u_1 and u_2 are the solutions of (1.1) satisfying the initial conditions

$$u_1(0) = 1, \ u_1'(0) = 0,$$
$$u_2(0) = 0, \ u_2'(0) = 1.$$

Indeed, a monodromy matrix for the $n \times n$ system

$$x'(t) = A(t)x(t), \quad A(t+T) = A(t), \quad t \in \mathbb{R}, \quad T > 0, \tag{1.6}$$

related to the fundamental matrix solution $X(t)$ of (1.6) is a nonsingular matrix C satisfying that

$$X(t+T) = X(t)C, \quad t \in \mathbb{R}.$$

Two monodromy matrices of (1.6) are similar (recall that matrices C_1 and C_2 are called similar if there exists a nonsingular matrix P such that $C_2 = P^{-1}C_1P$), because if $Y(t)$ is another fundamental solution then there exists a nonsingular matrix P such that

$$Y(t) = X(t)P,$$

and therefore

$$Y(t + T) = X(t + T)P = X(t)CP = Y(t)P^{-1}CP.$$

So, the monodromy matrix related to $Y(t)$ is $P^{-1}CP$.

Since the eigenvalues of two similar matrices are the same, we can define the characteristic multipliers of (1.6) as the eigenvalues $\rho \in \mathbb{C}$ of any monodromy matrix and a characteristic exponent of (1.6) as any $\alpha \in \mathbb{C}$ such that $\rho = e^{\alpha T}$. Note that the characteristic exponents are only uniquely defined modulo $2\pi i$. Moreover, if ρ_j, $j = 1, 2, \ldots, n$ are the characteristic multipliers of (1.6) it is well-known that, see [3],

$$\prod_{j=1}^{n} \rho_j = e^{\int_0^T tr(A(s))\,ds}. \tag{1.7}$$

Hill's equation (1.1) is equivalent to the system

$$\begin{pmatrix} u' \\ v' \end{pmatrix} = \begin{pmatrix} 0 & 1 \\ -a(t) & 0 \end{pmatrix} \begin{pmatrix} u \\ v \end{pmatrix} \tag{1.8}$$

which is a particular case of (1.6). As monodromy matrix we can choose $M(T)$, which is related to the principal matrix $M(t)$ defined in (1.5). Then, the characteristic multipliers of (1.1) are the eigenvalues of $M(T)$, that is, the roots of the characteristic equation

$$\rho^2 - (u_1(T) + u'_2(T))\,\rho + 1 = 0,$$

which by (1.7) satisfy $\rho_1\rho_2 = 1$. Therefore ρ_1 and ρ_2 are either real or complex conjugate numbers on the unit circle.

Another way to see $M(T)$ is as the matrix of the Poincaré map of (1.8): that is, if $(u_0, v_0) \in \mathbb{R}^2$ then the value of the solution of (1.8) through the initial condition (u_0, v_0) after a period T is given by

$$\begin{pmatrix} u(T) \\ v(T) \end{pmatrix} = M(T) \begin{pmatrix} u_0 \\ v_0 \end{pmatrix}.$$

A nonsingular matrix P is called stable if there exists a constant $c > 0$ such that $\|P^n\| < c$ for all $n \in \mathbb{Z}$, where $\| \cdot \|$ denotes any matrix norm. Then it is easy to show that (1.1) is stable if and only if its Floquet matrix $M(T)$ is stable.

In next section we will see in detail that the stability of the Hill's equation (1.1) is equivalent to $\rho_1 \neq \rho_2$ be complex conjugate on the unit circle or $\rho_1 = \rho_2 = \pm 1$ be an eigenvalue with geometric multiplicity two.

1.3 FLOQUET'S THEOREM FOR THE HILL'S EQUATION

Consider the Hill's equation in its standard form

$$u''(t) + (a(t) + \lambda)\, u(t) = 0, \quad \text{a.e. } t \in \mathbb{R}, \tag{1.9}$$

with $\lambda \in \mathbb{C}$ and $a \in L^{\infty}(\mathbb{R})$ a T-periodic function (we could also consider a function defined on the interval $[0,\ T]$ and take a as its periodic extension). In this section we will follow the line of [9].

It must be pointed out that all the theory developed from now on is valid for every closed interval $[a, b]$ and not only for $[0, T]$.

Definition 4. Under the previous conditions, Eq. (1.9) has two solutions, u_1 and u_2, which are uniquely determined by the following initial conditions:

$$u_1(0) = 1, \ u_1'(0) = 0,$$
$$u_2(0) = 0, \ u_2'(0) = 1.$$

These solutions are known as normalized solutions. In order to emphasize its dependence on the parameter λ, sometimes we will denote them by $u_1(t, \lambda)$ and $u_2(t, \lambda)$.

Definition 5. Given two solutions y_1, y_2 of Eq. (1.9) we define its Wronskian as

$$W(y_1, y_2)(t) := \det \begin{pmatrix} y_1(t) & y_2(t) \\ y_1'(t) & y_2'(t) \end{pmatrix} = y_1(t)y_2'(t) - y_1'(t)y_2(t).$$

Proposition 1. *The following claims hold:*
i) *If y_1, y_2 are solutions of (1.9) then its Wronskian $W(y_1, y_2)$ is constant.*
ii) *If u_1, u_2 are the normalized solutions given in Definition 4 then*

$$W(u_1, u_2)(t) = 1 \quad \text{for all } t \in \mathbb{R}.$$

Proof. If y_1, y_2 are solutions of (1.9) then we have that for a.e. $t \in \mathbb{R}$

$$\frac{d}{dt} W(y_1, y_2)(t) = y_1'(t)y_2'(t) + y_1(t)y_2''(t) - y_1''(t)y_2(t) - y_1'(t)y_2'(t)$$

$$= -\gamma_1(t)(a(t)+\lambda)\gamma_2(t) + (a(t)+\lambda)\gamma_1(t)\gamma_2(t) = 0.$$

Thus, since $W(\gamma_1,\gamma_2)$ is an absolutely continuous function, we have that it is constant. Now, from $W(u_1,u_2)(0) = 1$ the second claim also follows. □

Definition 6. The equation

$$\rho^2 - (u_1(T)+u_2'(T))\rho + 1 = 0 \qquad (1.10)$$

is known as the characteristic equation associated to (1.9) and we will denote its roots by ρ_1 and ρ_2, with $\rho_1,\rho_2 \in \mathbb{C}$.

Remark 1. Notice that the roots of the characteristic equation are the eigenvalues of the monodromy matrix

$$M(T) := \begin{pmatrix} u_1(T) & u_2(T) \\ u_1'(T) & u_2'(T) \end{pmatrix},$$

since by Proposition 1, ii) we know that $u_1(T)u_2'(T) - u_1'(T)u_2(T) = W(u_1,u_2)(T) = 1$.

On the other hand, since ρ_1 and ρ_2 are the roots of the characteristic equation, it is satisfied that

$$\rho^2 - (u_1(T)+u_2'(T))\rho + 1 = (\rho-\rho_1)(\rho-\rho_2) = \rho^2 - (\rho_1+\rho_2)\rho + \rho_1\rho_2,$$

from where we deduce that $\rho_1\rho_2 = 1$ and $\rho_1+\rho_2 = u_1(T)+u_2'(T)$.

Definition 7. Characteristic exponent refers to a complex number α satisfying that

$$e^{i\alpha T} = \rho_1, \qquad e^{-i\alpha T} = \rho_2.$$

Remark 2. The existence of such α is assured as a consequence of the equality $\rho_1\rho_2 = 1$. Moreover, if $\alpha = a+bi$, then

$$\rho_1 + \rho_2 = e^{i\alpha T} + e^{-i\alpha T} = e^{-bT}(\cos(aT)+i\sin(aT))$$
$$+ e^{bT}(\cos(-aT)+i\sin(-aT))$$
$$= \cos(aT)\left(e^{-bT}+e^{bT}\right) - i\sin(aT)\left(e^{bT}-e^{-bT}\right)$$
$$= 2(\cos(aT)\cosh(bT) - i\sin(aT)\sinh(bT))$$
$$= 2\cos(\alpha T).$$

Theorem 2 (Floquet). *If $\rho_1 \neq \rho_2$, then the Hill's equation (1.9) has two linearly independent solutions*

$$f_1(t) = e^{i\alpha t} p_1(t) \quad and \quad f_2(t) = e^{-i\alpha t} p_2(t),$$

with p_1 and p_2 T-periodic functions and $\alpha \in \mathbb{C}$ given in Definition 7.

If $\rho_1 = \rho_2$, then Hill's equation has a solution which is T-periodic (if $\rho_1 = \rho_2 = 1$) or T-antiperiodic (if $\rho_1 = \rho_2 = -1$). If we denote this solution by p and u is any other linearly independent solution, then it is satisfied that

$$u(t + T) = \rho_1 u(t) + \theta p(t)$$

for some constant θ. Moreover, condition $\theta = 0$ is equivalent to

$$u_1(T) + u_2'(T) = \pm 2, \quad u_2(T) = 0, \quad u_1'(T) = 0.$$

Proof. We consider two different cases.

Case 1: $\rho_1 \neq \rho_2$.

If u is a solution of (1.9) then, since $a(t) = a(t + T)$, $u(t + T)$ is also a solution. Consequently, $u_1(t + T)$ and $u_2(t + T)$ are solutions of (1.9), with u_1 and u_2 given in Definition 4.

Since u_1 and u_2 are linearly independent, every solution of (1.9) can be expressed as a linear combination of them. In particular, we have that

$$u_1(t + T) = u_1(T) u_1(t) + u_1'(T) u_2(t) \tag{1.11}$$

and

$$u_2(t + T) = u_2(T) u_1(t) + u_2'(T) u_2(t). \tag{1.12}$$

Suppose now that there exists a nontrivial solution of (1.9) (which we will denote by u) verifying that

$$u(t + T) = \rho u(t)$$

for some constant $\rho \in \mathbb{C}$.

As $u(t) = c_1 u_1(t) + c_2 u_2(t)$ for certain $c_1, c_2 \in \mathbb{C}$, then

$$u(t + T) = c_1 \left(u_1(T) u_1(t) + u_1'(T) u_2(t) \right) + c_2 \left(u_2(T) u_1(t) + u_2'(T) u_2(t) \right).$$

Thus, it will happen that $u(t + T) = \rho u(t) = \rho c_1 u_1(t) + \rho c_2 u_2(t)$ if and only if the parameters c_1 and c_2 satisfy the following linear algebraic system

$$(u_1(T) - \rho) c_1 + u_2(T) c_2 = 0, \tag{1.13}$$

$$u_1'(T)c_1 + (u_2'(T) - \rho)c_2 = 0. \tag{1.14}$$

A necessary and sufficient condition for the previous system to have a non-trivial solution (c_1, c_2) is that

$$\begin{vmatrix} u_1(T) - \rho & u_2(T) \\ u_1'(T) & u_2'(T) - \rho \end{vmatrix} = \rho^2 - (u_1(T) + u_2'(T))\rho + u_1(T)u_2'(T)$$
$$- u_2(T)u_1'(T) = 0.$$

Since the Wronskian $u_1(t)u_2'(t) - u_2(t)u_1'(t) = 1$ for all t, the previous equation is equivalent to (1.10). Then, if $\rho = \rho_1$ is a root of (1.10), the existence of c_1 and c_2 such that $u = c_1 u_1 + c_2 u_2 \neq 0$ satisfies $u(t + T) = \rho_1 u(t)$ is guaranteed. We can write then

$$u(t) = e^{i\alpha t}p_1(t) = f_1(t),$$

where $e^{i\alpha T} = \rho_1$ and p_1 is a T-periodic function.

Analogously, if the characteristic equation has another root $\rho_2 \neq \rho_1$, we can construct another solution

$$f_2(t) = e^{-i\alpha t}p_2(t)$$

with $e^{-i\alpha T} = \rho_2$ and p_2 a T-periodic function.

Both solutions are linearly independent since, on the contrary, there would exist nonzero constants $\lambda_1, \lambda_2 \in \mathbb{C}$ such that

$$\lambda_1 f_1(t) + \lambda_2 f_2(t) = 0, \quad \forall t \in \mathbb{R}.$$

In particular, it would be satisfied that

$$0 = \lambda_1 f_1(t + T) + \lambda_2 f_2(t + T) = \lambda_1 \rho_1 f_1(t) + \lambda_2 \rho_2 f_2(t)$$
$$= \rho_1 (\lambda_1 f_1(t) + \lambda_2 f_2(t)) + \lambda_2 (\rho_2 - \rho_1)f_2(t) = \lambda_2 (\rho_2 - \rho_1)f_2(t), \quad \forall t \in \mathbb{R},$$

which is only possible when $\rho_1 = \rho_2$.

So Floquet's Theorem is proved in the case that $\rho_1 \neq \rho_2$.

Case 2: $\rho_1 = \rho_2$.

In this case, it is still possible to construct a solution u_1^* of (1.9) verifying that $u_1^*(t + T) = \rho_1 u_1^*(t)$. On the other hand, as $\rho_1 \rho_2 = 1$, then necessarily $\rho_1 = \pm 1$. If $\rho_1 = 1$, then u_1^* is T-periodic. On the contrary, if $\rho_1 = -1$, since $u_1^*(t + T) = -u_1^*(t)$, u_1^* is T-antiperiodic.

We will find now another solution u_2^* that is linearly independent with u_1^*. First of all, consider the case in which $u_2(T) \neq 0$. Then we can choose

$$u_1^*(t) = u_2(T) u_1(t) + (\rho_1 - u_1(T)) u_2(t)$$

and

$$u_2^*(t) = u_2(t).$$

Observe that u_1^* verifies that $u_1^*(t + T) = \rho_1 u_1^*(t)$ because the constants $c_1 = u_2(T)$ and $c_2 = \rho_1 - u_1(T)$ constitute a nontrivial solution of system (1.13)–(1.14) (for $\rho = \rho_1$). Moreover, since $2\rho_1 = u_1(T) + u_2'(T)$, using (1.12), the following equality holds

$$u_2^*(t + T) = u_2(T) u_1(t) + (\rho_1 - u_1(T)) u_2(t) + \rho_1 u_2(t) = \rho_1 u_2^*(t) + u_1^*(t),$$

where, with the notation of the theorem, $\theta = 1$.

Analogously, if $u_2(T) = 0$, we could take

$$u_1^*(t) = u_2(t)$$

and

$$u_2^*(t) = u_1(t).$$

As $u_1(T) u_2'(T) - u_1'(T) u_2(T) = 1$, $u_2(T) = 0$, and $u_1(T) + u_2'(T) = 2\rho_1$, we deduce that $u_1(T) = u_2'(T) = \rho_1$ and so, from (1.11) and (1.12),

$$u_1^*(t + T) = \rho_1 u_1^*(t),$$

and

$$u_2^*(t + T) = \rho_1 u_2^*(t) + u_1'(T) u_1^*(t),$$

considering now $\theta = u_1'(T)$.

As a consequence, we deduce that condition $\theta = 0$ is equivalent to

$$u_1(T) + u_2'(T) = \pm 2, \quad u_2(T) = 0, \quad u_1'(T) = 0. \qquad \square$$

Remark 3. With the notation used in the previous proof, if $u = \gamma u_1^* + \beta u_2^*$ is any solution of Eq. (1.9), it satisfies

$$u(t + T) = \gamma \rho_1 u_1^*(t) + \beta \rho_1 u_2^*(t) + \beta \theta u_1^*(t) = \rho_1 u(t) + \beta \theta u_1^*(t).$$

Corollary 1. *If Eq. (1.9) has a nontrivial and nT-periodic solution for some $n \in \mathbb{N}$, $n > 2$, but neither T-periodic nor T-antiperiodic solutions, then all the solutions of (1.9) are periodic with period nT.*

Proof. The hypothesis that Eq. (1.9) has neither T-periodic nor T-antiperiodic solutions implies that we are in the case in which $\rho_1 \neq \rho_2$, so Floquet's Theorem warrants the existence of two linearly independent solutions, f_1 and f_2, such that

$$f_1(t) = e^{i \alpha t} p_1(t),$$

and

$$f_2(t) = e^{-i \alpha t} p_2(t),$$

with $\alpha \in \mathbb{C}$, $e^{i \alpha T} = \rho_1$, $e^{-i \alpha T} = \rho_2$, and p_1, and p_2 T-periodic functions.

Let u be a nontrivial and nT-periodic solution of (1.9), which can be expressed as a linear combination of f_1 and f_2:

$$u = \mu f_1 + \sigma f_2.$$

As a consequence, it is satisfied that

$$u(t) = u(t + nT) = \mu c f_1(t) + \sigma c' f_2(t),$$

with $c = e^{i \alpha n T}$ and $c' = e^{-i \alpha n T}$.

Since f_1 and f_2 are linearly independent, necessarily $c = c' = 1$. Therefore, if $\alpha = a + bi$ we obtain

$$1 = e^{i \alpha n T} = e^{i a n T} e^{-b n T} = e^{-b n T} (\cos(a n T) + i \sin(a n T))$$

and

$$1 = e^{-i \alpha n T} = e^{-i a n T} e^{b n T} = e^{b n T} (\cos(a n T) - i \sin(a n T)),$$

from where we deduce that $b = 0$ and $a n T = 2k\pi$ for some $k \in \mathbb{Z}$. Thus,

$$f_1(t) = e^{i \frac{2k\pi}{nT} t} p_1(t) \text{ and } f_2(t) = e^{-i \frac{2k\pi}{nT} t} p_2(t)$$

are nT-periodic, and the same occurs to any other solution of (1.9).   □

Remark 4. If $\rho_1 \neq \rho_2$, every nontrivial solution of (1.9) can be expressed as a linear combination of $f_1(t) = e^{i \alpha t} p_1(t)$ and $f_2(t) = e^{-i \alpha t} p_2(t)$. Therefore, if α takes a real value, every solution of (1.9) has an upper bound in absolute value which does not depend on t. On the contrary, if α is not real, we can assure the existence of a nontrivial and unbounded solution of (1.9).

On the other hand, if $\rho_1 = \rho_2 = \pm 1$, every nontrivial solution of (1.9) is a linear combination of a periodic or antiperiodic function p and another function u verifying that $u(t + T) = \rho_1 u(t) + \theta\, p(t)$. Consequently, all the solutions of (1.9) are bounded if and only if $\theta = 0$ or, equivalently, $u_1(T) + u_2'(T) = \pm 2$, $u_2(T) = 0$ and $u_1'(T) = 0$.

Taking into account the previous remark, we have the following characterization of stability.

Corollary 2. *The trivial solution of* (1.9) *is stable if and only if we are in one of the following situations:*

1. $u_1(T) + u_2'(T)$ *takes a real value and* $|u_1(T) + u_2'(T)| < 2$.
2. $u_1(T) + u_2'(T) = \pm 2$, $u_2(T) = u_1'(T) = 0$.

Proof. First case corresponds with $\rho_1 \neq \rho_2$ and $\alpha \in \mathbb{R}$. In fact, by Remarks 1 and 2, $u_1(T) + u_2'(T) = \rho_1 + \rho_2 = 2\cos(\alpha\, T)$ and this quantity is real if and only if α is real or $\alpha = \frac{k\pi}{T} + bi$, with $b \neq 0$. In this last case we would have that $|u_1(T) + u_2'(T)| > 2$. On the other hand, if α is real, $|u_1(T) + u_2'(T)| \leq 2$ and the equality holds when $\alpha\, T = k\pi$, with $k \in \mathbb{Z}$, which is not possible since, in that case, it would occur that $\rho_1 = e^{ik\pi} = e^{-ik\pi} = \rho_2$.

Second case corresponds with $\rho_1 = \rho_2$ and $\theta = 0$. $\square$

Remark 5. Notice that Condition 1 of Corollary 2 is equivalent to say that the characteristic multipliers $\rho_1 \neq \rho_2$ are complex conjugate numbers on the unit circle.

On the other hand, Condition 2 of Corollary 2 is equivalent to $\rho_1 = \rho_2 = \pm 1$ and that all the solutions of (1.9) are periodic or antiperiodic, respectively.

Example 1. In general one can not expect to know explicitly the discriminant $u_1(T) + u_2'(T)$ of Hill's equation. One exception is the equation introduced by Meissner in 1918, see [10], where the potential a is a piecewise constant function. An elementary analysis of its stability is given in [7] and see also [1] for an interesting related problem.

If we consider the simplest case

$$a(t) = \begin{cases} c_1^2, & \text{if } 0 \leq t \leq \pi, \\ c_2^2, & \text{if } \pi < t \leq 2\pi, \end{cases}$$

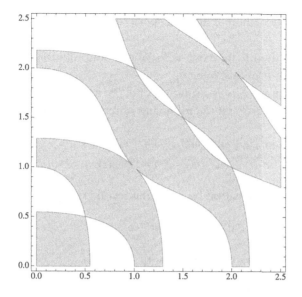

Figure 1.1 Stability region for Meissner's equation.

with $c_1, c_2 > 0$, it is easy to obtain the explicit expression of the discriminant by direct computation

$$u_1(2\pi) + u_2'(2\pi) = 2\cos(c_1\pi)\cos(c_2\pi) - \frac{\left(c_1^2 + c_2^2\right)\sin(c_1\pi)\sin(c_2\pi)}{c_1 c_2}.$$

The stability region in the $c_1 c_2$-plane given by condition $|u_1(2\pi) + u_2'(2\pi)| < 2$ is sketched in Fig. 1.1.

Theorem 3. *The trivial solution of* (1.9) *is unstable when* λ *takes a complex value.*

Proof. Let $\lambda = \mu + i\nu$, with $\nu \neq 0$, and suppose that the trivial solution of (1.9) is stable. Then, Floquet's Theorem warrants the existence of a solution of the form

$$y(t) = u(t) + i\nu(t) = e^{i\alpha t}p(t),$$

with $\alpha \in \mathbb{R}$ the characteristic exponent and p a T-periodic function. Indeed:

- If $\rho_1 \neq \rho_2$, then there exists $f_1(t) = e^{i\alpha t}p_1(t)$, where p_1 is T-periodic. Since the trivial solution is stable, α needs to be real, as we have noted in Remark 4.

- If $\rho_1 = \rho_2 = 1$, then there exists a T-periodic solution p (in this case $\alpha = 0$).
- If $\rho_1 = \rho_2 = -1$, then there exists a T-antiperiodic solution, so we can take $\alpha = \frac{k\pi}{T}$, with k an odd integer.

Therefore, since $y = u + iv$ is a solution of (1.9), it can be separated into its real and imaginary parts and the following system is fulfilled

$$u''(t) + (\mu + a(t))\,u(t) = v\,v(t),$$
$$v''(t) + (\mu + a(t))\,v(t) = -v\,u(t).$$

Multiplying the first equation by v and the second one by u, we reach

$$u''\,v - v''\,u = v\left(u^2 + v^2\right)$$

and integrating previous expression on $[0, t]$ we obtain

$$u'(t)\,v(t) - v'(t)\,u(t) = v\int_0^t \left(u^2(x) + v^2(x)\right)\,dx + k.$$

As $y(t) = e^{i\alpha t}p(t) = u(t) + iv(t)$, u, $v \in \mathcal{C}^1(\mathbb{R})$ and u, v, u', v' are T-periodic, then $|u|$, $|v|$, $|u'|$, and $|v'|$ are bounded on $\mathbb{R}$ and, consequently, there exists an upper bound for $|u\,v' - v\,u'|$ which is independent of t. Therefore

$$|I(t)| = \left|\int_0^t \left(u^2(x) + v^2(x)\right)\,dx\right|$$

also has an upper bound which does not depend on t.

However, $u^2 + v^2 = |p|^2$ and then

$$I(n\,T) = \int_0^{nT} \left(u^2(x) + v^2(x)\right)\,dx = n\int_0^T |p(x)|^2\,dx$$

for $n \in \mathbb{N}$, which implies that $|I(t)| \xrightarrow[t\to\infty]{} \infty$.

We reach a contradiction and thus we conclude that the trivial solution is not stable for (1.9) when λ takes a complex value. $\qquad\square$

Theorem 4. *There exists $\lambda^* \in \mathbb{R}$ verifying that for all $\lambda \le \lambda^*$, the trivial solution of (1.9) is unstable.*

Proof. Let λ^* be such that $\lambda^* + a(t) < 0$ for a.e. $t \in [0, T]$. We know that this value λ^* exists because a is essentially bounded. Let us see that $u_1(t, \lambda) \xrightarrow[t\to\infty]{} \infty$ for all $\lambda \le \lambda^*$.

Since $\lambda^* + a(t) < 0$ for a.e. $t \in [0, T]$, u_1 and u_1'' have the same sign. As $u_1(0) = 1 > 0$, then $u_1'' > 0$ almost everywhere in a neighbourhood of 0 and u_1' is strictly increasing on such neighbourhood. Since $u_1'(0) = 0$, we deduce that there exists some $\delta > 0$ such that $u_1'(t) > 0$ for $t \in (0, \delta)$. As a consequence, u_1 is strictly increasing (and then $u_1 > 1$) on $(0, \delta)$. We can continue with this argument in all the positive axis, showing that $u_1 > 1$ on $(0, \infty)$ and that $\lim_{t \to \infty} u_1(t, \lambda) = \infty$.

Using a similar argument, it could be proved that $u_2' > 1$ on $(0, \infty)$ and so $\lim_{t \to \infty} u_2(t, \lambda) = \infty$.   □

From the proof of the previous theorem we deduce the following.

Lemma 1. *If* $\lambda \leq \lambda^*$, *then* $u_1(t, \lambda) + u_2'(t, \lambda) > 2$ *for all* $t > 0$.

Since the computation of the characteristic equation (1.10) is not easy, because we have to know the normalized solutions u_1 and u_2, some sufficient conditions for the stability of (1.1) given explicitly in terms on the potential a have been developed, being Lyapunov's stability criterium the most famous one.

Theorem 5 (Lyapunov's stability criterium). *If* $a \not\equiv 0$, $\int_0^T a(s)\,ds \geq 0$ *and*

$$\int_0^T |a(s)|\,ds < \frac{4}{T}, \tag{1.15}$$

then the trivial solution of Eq. (1.1) is stable.

Proof. We are going to discard the existence of real solutions of the characteristic equation in order to ensure stability. To the contrary suppose that $\rho \in \mathbb{R}$ is an eigenvalue of $M(T)$. Then, since $M(t + T) = M(t)M(T)$, we have that there exists a solution u of (1.1) such that $u(t + T) = \rho u(t)$. Thus, either $u(t) \neq 0$ for all $t \in \mathbb{R}$ or u has two consecutive zeros t_0, t_1 with $0 < t_1 - t_0 \leq T$. The last is impossible because of (1.15) and the Lyapunov inequality, see [11], which implies that

$$\int_0^T |a(s)|\,ds \geq \int_{t_1}^{t_2} |a(s)|\,ds \geq \frac{4}{t_1 - t_0} \geq \frac{4}{T}.$$

The former possibility also leads to a contradiction because dividing (1.1) by $u(t)$, integrating by parts between 0 and T, and taking into account that

$$\frac{u'(T)}{u(T)} = \frac{\rho u'(0)}{\rho u(0)} = \frac{u'(0)}{u(0)},$$

we obtain that

$$\int_0^T \frac{u''(s)}{u(s)}\, ds + \int_0^T a(s)\, ds = \int_0^T \frac{(u')^2(s)}{u^2(s)}\, ds + \int_0^T a(s)\, ds = 0.$$

Therefore u should be constant, which is a contradiction because $a \not\equiv 0$. Hence, the solutions $\rho_1 \neq \rho_2$ of the characteristic equation are complex conjugate numbers on the unit circle and by Remark 5 the trivial solution of Eq. (1.1) is stable. $\square$

Further results concerning Lyapunov's stability criterium and its generalizations can be found in [3]. See also [11] for some extensions of the Lyapunov inequality.

REFERENCES

[1] J.M. Almira, P.J. Torres, Invariance of the stability of Meissner's equation under a permutation of its intervals, Ann. Mat. Pura Appl. (4) 180 (2001) 245–253.
[2] A. Cabada, J.A. Cid, On the sign of the Green's function associated to Hill's equation with an indefinite potential, Appl. Math. Comput. 205 (2008) 303–308.
[3] L. Cesari, Asymptotic Behavior and Stability Problems in Ordinary Differential Equations, third edition, Springer, 1970.
[4] L. Csizmadia, L. Hatvani, An Extension of the Levi-Weckesser Method to the Stabilization of the Inverted Pendulum Under Gravity, Springer Science+Business Media Dordrecht, 2013.
[5] J.K. Hale, Ordinary Differential Equations, second edition, Robert E. Krieger Publishing Co., New York, 1980.
[6] G.W. Hill, On the part of the motion of the lunar perigee which is a function of the mean motions of the Sun and the Moon, Acta Math. VIII (1886) 1–36.
[7] H. Hochstadt, A special Hill's equation with discontinuous coefficients, Amer. Math. Monthly 70 (1963) 18–26.
[8] M. Levi, W. Weckesser, Stabilization of the inverted linearized pendulum by high frequency vibrations, SIAM Rev. 37 (1995) 219–223.
[9] W. Magnus, S. Winkler, Hill's Equation, Dover Publications, New York, 1979.
[10] E. Meissner, Ueber Schuettelschwingungen in Systemen mit periodisch veraenderlicher Elastizitaet, Schweizer Bauzeitung, vol. 72, 1918, pp. 95–98.
[11] J.P. Pinasco, Lyapunov-Type Inequalities, Springer Briefs Math., 2013.
[12] N. Rouche, J. Mawhin, Ordinary Differential Equations, Surveys and Reference Works in Mathematics, Pitman, Boston, 1980.
[13] G.F. Simmons, J.S. Robertson, Differential Equations With Applications and Historical Notes, McGraw-Hill, 1991.
[14] P.J. Torres, Existence of one-signed periodic solutions of some second-order differential equations via a Krasnoselskii fixed point theorem, J. Differential Equations 190 (2003) 643–662.
[15] M. Zhang, W. Li, A Lyapunov-type stability criterion using L^α norms, Proc. Amer. Math. Soc. 130 (2002) 3325–3333.

CHAPTER 2

Homogeneous Equation

Contents

2.1 Introduction   19
2.2 Sturm Comparison Theory   20
2.3 Spectral Properties of Dirichlet Problem   26
2.4 Spectral Properties of Mixed and Neumann Problems   29
2.5 Spectral Properties of the Periodic Problem: Intervals of Stability and Instability   31
2.6 Relation Between Eigenvalues of Neumann, Dirichlet, Periodic, and
Antiperiodic Problems   42
References   47

2.1 INTRODUCTION

Consider the homogeneous linear differential equation of second order

$$y''(t) + p(t)\, y'(t) + q(t)\, y(t) = 0, \quad \text{a.e. } t \in \mathbb{R} \tag{2.1}$$

with $p, q \in L^\infty(\mathbb{R})$.

This equation is not necessarily solvable in terms of elementary functions. However, it is possible to establish some qualitative properties of its solutions. In particular, we will be interested in studying their oscillation, for which we will make several previous considerations.

In particular, we will show that, under certain regularity conditions, the term containing the first derivative on Eq. (2.1) can be eliminated through a change of variable that does not modify the oscillation properties of the solutions. This will lead to a Hill's equation

$$u''(t) + a(t)\, u(t) = 0, \quad \text{a.e. } t \in \mathbb{R}.$$

All along the chapter, we will work in the space $W^{2,\infty}_{loc}(\mathbb{R})$, defined as the set of functions $y \in \mathcal{C}^1(\mathbb{R})$ such that y' is absolutely continuous on $\mathbb{R}$ and $y'' \in L^\infty_{loc}(\mathbb{R})$.

This chapter is organized as follows: Section 2.2, following the line of [5], gives a general description of the oscillation properties of the solutions. Section 2.3 characterizes the spectrum of the Dirichlet problem

$$u''(t) + (a(t) + \lambda)\, u(t) = 0, \quad \text{a.e. } t \in [0,\, T] \equiv I, \quad u(0) = u(T) = 0,$$

Maximum Principles for the Hill's Equation.
DOI: http://dx.doi.org/10.1016/B978-0-12-804117-8.00002-3
Copyright © 2018 Elsevier Inc. All rights reserved.

while at Section 2.4 we deal with the spectrum of Neumann

$$u''(t) + (a(t) + \lambda)\, u(t) = 0, \quad \text{a.e. } t \in [0, T], \quad u'(0) = u'(T) = 0$$

and mixed problems

$$u''(t) + (a(t) + \lambda)\, u(t) = 0, \quad \text{a.e. } t \in [0, T], \quad u'(0) = u(T) = 0,$$
$$u''(t) + (a(t) + \lambda)\, u(t) = 0, \quad \text{a.e. } t \in [0, T], \quad u(0) = u'(T) = 0.$$

The main result is obtained in Section 2.5, where it is studied the periodic problem

$$u''(t) + (a(t) + \lambda)\, u(t) = 0, \quad \text{a.e. } t \in [0, T], \quad u(0) = u(T),\ u'(0) = u'(T),$$

and the antiperiodic one

$$u''(t) + (a(t) + \lambda)\, u(t) = 0, \quad \text{a.e. } t \in [0, T], \quad u(0) = -u(T),\ u'(0) = -u'(T),$$

characterizing their spectrum and relating them with the intervals of stability and instability of the Hill's equation. This section follows the line of [4], introducing some new details in several proofs and simplifying the one given in that reference for the Oscillation Theorem. Some of these results can also be found in [3]. Finally in Section 2.6 we present a relationship between the order of the eigenvalues of Neumann, Dirichlet, periodic, and antiperiodic problems.

2.2 STURM COMPARISON THEORY

We will study now the Sturm comparison theory, establishing the main general properties of oscillation of solutions.

Theorem 6 (Sturm's separation). *Let y_1 and y_2 be two linearly independent solutions of (2.1). Then, neither y_1 and y_2 nor y_1' and y_2' can have any zero in common. Moreover, y_1 vanishes exactly once between two consecutive zeros of y_2, and reciprocally.*

Proof. Let y_1 and y_2 be in the conditions of the theorem. Since they are linearly independent solutions of the equation, its Wronskian

$$W(y_1, y_2) = y_1\, y_2' - y_2\, y_1'$$

is different from zero on $\mathbb{R}$ and so, as it is continuous, it has constant sign.

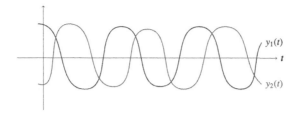

Figure 2.1 Two linearly independent solutions of (2.1).

As an immediate consequence we deduce that neither y_1 and y_2 nor y_1' and y_2' can have any zero in common, otherwise the Wronskian would be zero in that point.

Consider now t_1 and t_2 two consecutive zeros of y_2. In such points, the Wronskian takes the form $y_1\, y_2'$ and, since it is different from zero, both y_1 and y_2' are different from zero at t_1 and t_2. In addition, $y_2'(t_1)$ and $y_2'(t_2)$ take opposite signs.

As the Wronskian has constant sign, then $y_1(t_1)$ and $y_1(t_2)$ also have different signs, from where we deduce, given the continuity of y_1, that it must vanish somewhere on (t_1, t_2).

Moreover, y_1 has a unique zero between t_1 and t_2 since on the contrary the previous argument would imply that y_2 takes the value zero between two consecutive zeros of y_1, which contradicts the hypothesis that t_1 and t_2 are consecutive zeros of y_2. ☐

The situation described on the previous theorem would be similar to the one represented in Fig. 2.1.

We will simplify now expression (2.1), seeing that every equation in this form, in which p and q satisfy suitable regularity conditions, could be rewritten as a Hill's equation, also called the normal form of (2.1),

$$u''(t) + a(t)\, u(t) = 0, \quad \text{a.e. } t \in \mathbb{R}. \tag{2.2}$$

In order to write (2.1) in the normal form, we decompose $y(t) = u(t)\, v(t)$, so that $y' = uv' + u'v$ and $y'' = uv'' + 2u'v' + u''v$. Substituting in (2.1), we obtain

$$v u'' + (2v' + pv)\, u' + (v'' + pv' + qv)\, u = 0.$$

Making the coefficient of u' equal to zero, we deduce that, for some $t_0 \subset \mathbb{R}$,

$$v(t) = e^{-\frac{1}{2} \int_{t_0}^t p(s) \, ds}$$

reduces (2.1) into the normal form (2.2), with

$$a(t) = \frac{v''(t)}{v(t)} + p(t) \frac{v'(t)}{v(t)} + q(t).$$

Taking into account that

$$\frac{v'(t)}{v(t)} = -\frac{1}{2} p(t)$$

and

$$\frac{v''(t)}{v(t)} = -\frac{1}{2} p'(t) + \frac{1}{4} p^2(t),$$

we obtain

$$a(t) = q(t) - \frac{1}{4} p^2(t) - \frac{1}{2} p'(t).$$

We observe that, since v does not take the value zero, the transformation we have just made does not affect neither to the zeros of the solutions nor to their oscillation and sign.

Consequently, from now on, we will work with the Hill's equation (2.2). We will be specially interested in those cases in which either a is a positive function or it changes its sign. This is due to the following result.

Theorem 7. *If $a < 0$ a.e. on $\mathbb{R}$, every nontrivial solution of (2.2) has at most one zero on $\mathbb{R}$.*

Proof. Assume $a(t) < 0$ for a.e. $t \in \mathbb{R}$ in Eq. (2.2) and let u be a nontrivial solution.

If t_0 is a zero of u, since the solution is not trivial, necessarily $u'(t_0) \neq 0$. Suppose that $u'(t_0) > 0$. Then u takes positive values on some interval on the right side of t_0 and, consequently, $u''(t) = -u(t) a(t) > 0$ for a.e. t on that interval. Therefore, u' is an increasing function on the right of t_0, which implies that u does not have any zero on the right side of t_0. The same way, u cannot have zeros on the left side of t_0.

In case of being $u'(t_0) < 0$, the argument would be analogous.   □

Thus, as if $a < 0$ a.e. on $\mathbb{R}$ the solutions do not oscillate, we will not consider this case. Nevertheless, condition $a > 0$ a.e. on $\mathbb{R}$ does not assure the oscillation. In fact, let u be a nontrivial solution of (2.2) with $a > 0$ a.e. on $\mathbb{R}$. If we consider an interval on which $u > 0$, then $u''(t) = -a(t)\,u(t) < 0$ for a.e. $t \in \mathbb{R}$, that is, u' is decreasing. If this slope becomes negative, u will have a zero somewhere on the right side of the considered interval. However, if u' decreases but remains positive, then u is strictly increasing and never takes value zero.

We can find an example of this situation by considering Euler's equation

$$u''(x) + \frac{k}{x^2}\,u(x) = 0, \quad x > 0, \tag{2.3}$$

with k a positive constant.

The change of variable $t = \log(x)$ allows us to transform the previous equation into a homogeneous linear equation with constant coefficients. In fact, we have the following equalities

$$v(t) := u(e^t),$$
$$v'(t) = u'(e^t)\,e^t,$$
$$v''(t) = u''(e^t)\,e^{2t} + u'(e^t)\,e^t.$$

From (2.3), we reach the following equation with constant coefficients

$$v''(t) - v'(t) + k\,v(t) = 0.$$

This equation can be explicitly solved, its general solution follows the expression

- $v(t) = c_1\,e^{\frac{1+\sqrt{1-4k}}{2}\,t} + c_2\,e^{\frac{1-\sqrt{1-4k}}{2}\,t}$, if $k < \frac{1}{4}$,
- $v(t) = c_1\,e^{\frac{1}{2}t} + c_2\,t\,e^{\frac{1}{2}t}$, if $k = \frac{1}{4}$,
- $v(t) = e^{\frac{1}{2}t}\left(c_1 \cos \frac{\sqrt{4k-1}}{2}t + c_2 \sin \frac{\sqrt{4k-1}}{2}t\right)$, if $k > \frac{1}{4}$.

As a consequence, the solutions oscillate if and only if $k > \frac{1}{4}$.

In relation with what we have just commented, the following theorem gives a sufficient condition to assure the oscillation of the solutions.

Theorem 8. *Let u be a nontrivial solution of $u''(t) + a(t)\,u(t) = 0$, with $a \in L^{\infty}(\mathbb{R})$. Suppose that there exists $\bar{t} \in \mathbb{R}$ such that $a(t) > 0$ for a.e. $t > \bar{t}$ and*

$$\int_{\bar{t}}^{\infty} a(t)\,dt = \infty.$$

Then u has an infinite number of zeros bigger than $\bar{t}$.

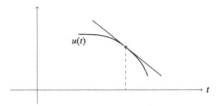

Figure 2.2 Graph of u for t large enough.

Proof. Assume that u vanishes on, at most, a finite number of points on $(\bar{t}, \infty)$. Then, there exists $t_0 > \bar{t}$ such that $u(t) \neq 0$ for $t \geq t_0$. Suppose that $u(t) > 0$ for all $t \geq t_0$. If we define $v(t) = -\frac{u'(t)}{u(t)}$, then $v'(t) = a(t) + v^2(t)$, and integrating from t_0 to t we obtain that

$$v(t) - v(t_0) = \int_{t_0}^{t} a(s)\, ds + \int_{t_0}^{t} v^2(s)\, ds.$$

As by hypothesis we have that $\int_{}^{\infty} a(t)\, dt = \infty$, then we can affirm that $v(t) > 0$ for t large enough and that, for such values of t, $u(t)$ and $u'(t)$ have opposite signs. In particular, $u'(t)$ would be negative. In addition, $u'' = -au < 0$ a.e. on (t_0, ∞), so u is concave and decreasing. See Fig. 2.2.

This implies that u has one zero on the right side of t_0. So we reach a contradiction and conclude that u must have an infinite number of zeros on the positive axis.

In case of being $u(t) < 0$ for $t \geq t_0$ the proof would be analogous.   □

Remark 6. The condition given in the previous theorem is sufficient but not necessary. To confirm this, it is enough to consider Eq. (2.3) defined on $(\bar{t}, \infty)$ for some $\bar{t} > 0$, for which we have that

$$\int_{\bar{t}}^{\infty} \frac{k}{x^2}\, dx = \frac{k}{\bar{t}}.$$

Nevertheless, as we have already commented, its solutions have an infinite number of zeros if and only if $k > \frac{1}{4}$.

Theorem 8 proves that, under certain conditions, a solution of (2.2) oscillates infinitely many times on an unbounded from above interval. The following theorem will prove that such solution can not oscillate infinitely many times on a compact interval and, consequently, its oscillations must extend through all the positive axis.

Theorem 9. *Let u be a nontrivial solution of equation $u''(t) + a(t) u(t) = 0$ on an interval $[c, d]$, with $a \in L^\infty([c, d])$. Then u has, at most, a finite number of zeros on such interval.*

Proof. Suppose that u has infinitely many zeros on the interval $[c, d]$. Then there exist a point $t_0 \in [c, d]$ and a sequence of zeros $\{t_n\}_{n \in \mathbb{N}}$, with $t_n \neq t_0$, such that $\lim_{n \to \infty} t_n = t_0$. Since $u \in \mathcal{C}^1[c, d]$, we have that

$$u(t_0) = \lim_{t_n \to t_0} u(t_n) = 0$$

and

$$u'(t_0) = \lim_{t_n \to t_0} \frac{u(t_n) - u(t_0)}{t_n - t_0} = 0.$$

As a consequence, $u'(t_0) = u(t_0) = 0$, and therefore u is the trivial solution. Thus, u can not have an infinite number of zeros on the interval $[c, d]$. □

Having seen this, we could think about comparing the way in which two different solutions of (2.2) oscillate. In this sense, Theorem 6 affirms that the zeros of every pair of nontrivial solutions of (2.2) either coincide or alternate, depending on their relation of linear dependence. Then, we can affirm that all the solutions of (2.2) oscillate with the same speed, in the sense that, on a given interval, two solutions will have the same number of zeros or one of them will have exactly one more zero than the other one.

On the other hand, the following theorem describes the influence of the potential on the speed of oscillation of the solutions.

Theorem 10 (Sturm's comparison). *Let u and v be nontrivial solutions of*

$$u''(t) + q(t) u(t) = 0, \quad a.e.\ t \in \mathbb{R}$$

and

$$v''(t) + r(t) v(t) = 0, \quad a.e.\ t \in \mathbb{R},$$

respectively, with $q, r \in L^1_{loc}(\mathbb{R})$ such that $q > r$ a.e. on $\mathbb{R}$. Then u vanishes at least once between two consecutive zeros of v.

Proof. Let t_1 and t_2 be two consecutive zeros of v and suppose that u does not vanish on the interval (t_1, t_2). By continuity, both u and v have constant sign on (t_1, t_2) and we will assume both are positive (in any other case the proof would be analogous).

Consider the Wronskian

$$W(u, v)(t) = u(t) v'(t) - v(t) u'(t).$$

Then

$$\frac{dW(u, v)}{dt} = u v'' - v u'' = u(-rv) - v(-qu) = (q - r) uv > 0$$

a.e. on the interval (t_1, t_2).

If we integrate the previous expression between t_1 and t_2 we obtain that $W(u, v)(t_2) > W(u, v)(t_1)$. However, since $v(t_1) = v(t_2) = 0$ and $v(t) > 0$ on (t_1, t_2), we have that $v'(t_1) \geq 0$ and $v'(t_2) \leq 0$ and so

$$W(u, v)(t_1) = u(t_1) v'(t_1) \geq 0$$

and

$$W(u, v)(t_2) = u(t_2) v'(t_2) \leq 0.$$

We reach a contradiction and we conclude that u attains a zero on the interval (t_1, t_2). $\square$

Remark 7. In particular, we can deduce from the previous theorem that, when $a(t) > k^2 > 0$ for a.e. $t \in \mathbb{R}$, every solution of Eq. (2.2) must have a zero between each pair of consecutive zeros of a solution $u(t) = \sin k (t - t_0)$ of the equation

$$u''(t) + k^2 u(t) = 0,$$

that is, it must have a zero on every interval of length $\frac{\pi}{k}$.

2.3 SPECTRAL PROPERTIES OF DIRICHLET PROBLEM

In this section, we will consider a different formulation of the problem which is equivalent to the one we have just seen: fix an interval $[0, T]$ and a function $a \in L^\infty([0, T])$ and consider the equation

$$u''(t) + (a(t) + \lambda) u(t) = 0, \quad \text{a.e. } t \in [0, T], \tag{2.4}$$

depending on the parameter $\lambda \in \mathbb{R}$.

If we extend periodically the function a to $\mathbb{R}$, it is trivially fulfilled that, for λ large enough, $a(t) + \lambda > 0$ for a.e. $t \in \mathbb{R}$ and

$$\int_0^\infty (a(t) + \lambda)\, dt = \infty$$

and so, as a consequence of Theorem 8, we know that every solution of (2.4) has an infinite number of zeros through the positive axis.

For such values of λ, $u_2(\cdot, \lambda)$ (see Definition 4) is the solution of (2.4) satisfying the initial conditions $u_2(0, \lambda) = 0$, $u_2'(0, \lambda) = 1$. Denote by $t_0(\lambda)$ the first positive zero of $u_2(\cdot, \lambda)$ (whose existence is warranted from Theorems 8 and 9). As a consequence of Theorem 10, we know that the bigger λ is, the smaller the value of $t_0(\lambda)$ is.

In particular, from Remark 7, we know that if λ is such that $\left(\frac{k\pi}{T}\right)^2 < a(t) + \lambda$ for a.e. $t \in [0, T]$, all the solutions of Eq. (2.4) vanish on every interval of length $\frac{T}{k}$, that is, they vanish at least k times on the interval $[0, T]$.

Consequently, if we take λ large enough, we can get the solutions $u_2(\cdot, \lambda)$ to have a finite number of zeros on the interval $[0, T]$ as big as we want.

Moreover, all these zeros must be simple since, on the contrary, $u_2(\cdot, \lambda) \equiv 0$, which would be a contradiction.

Remark 8. Notice that if λ verifies that

$$\left(\frac{k\pi}{T}\right)^2 < a(t) + \lambda < \left(\frac{(k+1)\pi}{T}\right)^2 \quad \text{for a.e. } t \in [0, T],$$

then $u_2(\cdot, \lambda)$ has exactly k zeros on the interval $[0, T]$.

Eq. (2.4) coupled with boundary value conditions

$$u(0) = u(T) = 0, \tag{2.5}$$

constitute the so-called Dirichlet problem.

Problem (2.4)–(2.5) has a nontrivial solution if and only if there exists some λ for which $u_2(T, \lambda) = 0$, $u_2(\cdot, \lambda) \neq 0$. Denote by S the set of values of λ satisfying such condition.

Theorem 7 warrants that if $a(t) + \lambda < 0$ for a.e. $t \in [0, T]$, problem (2.4)–(2.5) does not have a nontrivial solution. As a consequence, S is bounded from below.

Moreover, since $t_0(\lambda)$ decreases with respect to λ and converges to zero when λ goes to infinity, and taking into account the continuity of the problem with respect to the data, we deduce the existence of $\lambda_0^D[a]$ for which $t_0(\lambda_0^D[a]) = T$. Therefore, $\lambda_0^D[a] = \min S$.

Applying an analogous argument, we can keep on increasing λ until $u_2(\cdot, \lambda)$ has exactly three zeros on the interval $[0, T]$. Again, the continuity with respect to the data lets us conclude the existence of

$$\lambda_1^D[a] = \min\{\lambda > \lambda_0^D[a]; \ u_2(\cdot, \lambda) \text{ has exactly three zeros on } [0, T]\}.$$

Because of continuity, necessarily $u_2(T, \lambda_1^D[a]) = 0$. Then we have that

$$\lambda_1^D[a] = \min\left\{\lambda \in S \setminus \{\lambda_0^D[a]\}\right\}.$$

We could repeat this process an indefinite number of times, so we reach the conclusion that there exists a sequence of infinite values

$$\lambda_0^D[a] < \lambda_1^D[a] < \lambda_2^D[a] < \cdots < \lambda_k^D[a] < \cdots$$

for which problem (2.4)–(2.5), with $\lambda = \lambda_k^D[a]$, has a nontrivial solution $u_2(\cdot, \lambda_k^D[a])$ that has exactly k zeros on $(0, T)$. Note that, from Theorem 6, all the eigenvalues $\lambda = \lambda_k^D[a]$, $k = 0, 1, 2, \ldots$, are simple, that is, the associated eigenspace has dimension one.

Moreover, the previous sequence is unbounded from above. On the contrary, there would exist $n_0 \in \mathbb{N}$ such that $a(t) + \lambda_k^D[a] < \left(\frac{(n_0+1)\pi}{T}\right)^2$ for a.e. $t \in [0, T]$ and all $k = 0, 1, \ldots$. So, as a consequence of Theorem 10, all the solutions of problem (2.4)–(2.5), with $\lambda = \lambda_k^D[a]$, $k = 0, 1, \ldots$, would have at most n_0 zeros on $(0, T)$. This contradicts that $u_2(\cdot, \lambda_k^D[a])$ has exactly k zeros on $(0, T)$ for $k > n_0$. Consequently,

$$\lim_{k \to \infty} \lambda_k^D[a] = \infty.$$

Finally we note that, as a consequence of Sturm's comparison (Theorem 10), we could deduce that

Lemma 2. *The function*

$$\lambda_n^D \colon L^\infty(I) \to \mathbb{R},$$

$$a \mapsto \lambda_n^D[a],$$

that assigns to every potential a the n-th eigenvalue, $n = 0, 1, \ldots$, of the corresponding Dirichlet problem, is decreasing.

2.4 SPECTRAL PROPERTIES OF MIXED AND NEUMANN PROBLEMS

Consider now the following mixed problem (denoted as M_2)

$$u''(t) + (a(t) + \lambda) u(t) = 0, \quad \text{a.e. } t \in [0, T], \quad u(0) = u'(T) = 0. \tag{2.6}$$

It is clear that problem (2.6) has a nontrivial solution if and only if there exists some λ for which $u_2'(T, \lambda) = 0$, $u_2(\cdot, \lambda) \not\equiv 0$.

We have just proved in Section 2.3 that for λ large enough, $u_2(\cdot, \lambda)$ has a finite number of zeros on $[0, T]$ as big as we want. Since between two zeros of $u_2(\cdot, \lambda)$ there is at least one zero of $u_2'(\cdot, \lambda)$, it is clear that the number of zeros of $u_2'(\cdot, \lambda)$ on $[0, T]$ will also increase with λ.

Following the same argument that for Dirichlet problem, it is easy to deduce the existence of an infinite sequence of values

$$\lambda_0^{M_2}[a] < \lambda_1^{M_2}[a] < \lambda_2^{M_2}[a] < \cdots < \lambda_k^{M_2}[a] < \cdots$$

for which problem (2.6), with $\lambda = \lambda_k^{M_2}[a]$, has a nontrivial solution $u_2(\cdot, \lambda_k^{M_2}[a])$. Moreover, from Theorem 6, the eigenvalues $\lambda_k^{M_2}[a]$ for all $k = 0, 1, 2, \ldots$ are simple.

We will see now that $u_2(\cdot, \lambda_k^{M_2}[a])$ has exactly k zeros on $(0, T)$. First of all, since between two zeros of $u_2(\cdot, \lambda)$ there is at least one zero of $u_2'(\cdot, \lambda)$, it must occur that

$$\lambda_k^{M_2}[a] < \lambda_k^{D}[a]$$

for all k. The previous inequality is strict because all the zeros of $u_2(\cdot, \lambda)$ must be simple. As an immediate consequence, from Sturm's Comparison Theorem (Theorem 10) we deduce that $u_2(\cdot, \lambda_k^{M_2}[a])$ has at most k zeros on $(0, T)$.

Now we will see that it is not possible that two eigenfunctions related to consecutive eigenvalues of problem (2.6) have the same number of zeros on $(0, T)$. Assume, on the contrary, that both $u_2(\cdot, \lambda_k^{M_2}[a])$ and $u_2(\cdot, \lambda_{k+1}^{M_2}[a])$ have n zeros on $(0, T)$ and consider their even extensions ($\tilde{u}_2(\cdot, \lambda_k^{M_2}[a])$ and $\tilde{u}_2(\cdot, \lambda_{k+1}^{M_2}[a])$) to the interval $[0, 2T]$. It is clear that $\tilde{u}_2(\cdot, \lambda_k^{M_2}[a])$ and $\tilde{u}_2(\cdot, \lambda_{k+1}^{M_2}[a])$ solve the equation

$$u''(t) + (\tilde{a}(t) + \lambda) u(t) = 0, \quad \text{a.e. } t \in [0, 2T], \tag{2.7}$$

for $\lambda = \lambda_k^{M_2}[a]$ or $\lambda = \lambda_{k+1}^{M_2}[a]$, respectively, with $\tilde{a}$ the even extension of a to the interval $[0, 2T]$. Now, since $\tilde{u}_2(\cdot, \lambda_k^{M_2}[a])$ has $2n$ zeros on

$(0, 2T)$, $\tilde{u}_2(0, \lambda_k^{M_2}[a]) = \tilde{u}_2(2T, \lambda_k^{M_2}[a]) = 0$ and $\lambda_{k+1}^{M_2}[a] > \lambda_k^{M_2}[a]$, from Sturm's Comparison Theorem (Theorem 10) and since $\tilde{u}_2(T, \lambda_{k+1}^{M_2}[a]) \neq 0$ we deduce that $\tilde{u}_2(\cdot, \lambda_{k+1}^{M_2}[a])$ must have at least $2n + 1$ zeros on $(0, 2T)$, which is a contradiction.

As a consequence we deduce that $u_2(\cdot, \lambda_k^{M_2}[a])$ has at least k zeros on $(0, T)$ and, therefore, it has exactly k zeros on $(0, T)$.

Finally, it is easy to deduce that

$$\lambda_k^D[a] < \lambda_{k+1}^{M_2}[a].$$

Indeed, since $u_2(\cdot, \lambda_k^D[a])$ and $u_2(\cdot, \lambda_{k+1}^{M_2}[a])$ have k and $k+1$ zeros on $(0, T)$, respectively, the fact that $\lambda_{k+1}^{M_2}[a] < \lambda_k^D[a]$ would be a contradiction with Sturm's Comparison Theorem (Theorem 10).

Therefore, we deduce the following alternating property

$$\lambda_0^{M_2}[a] < \lambda_0^D[a] < \cdots < \lambda_k^{M_2}[a] < \lambda_k^D[a] < \lambda_{k+1}^{M_2}[a] < \lambda_{k+1}^D[a] < \ldots$$

Consider now Neumann problem (N)

$$u''(t) + (a(t) + \lambda)\,u(t) = 0, \quad \text{a.e. } t \in [0, T], \quad u'(0) = u'(T) = 0, \qquad (2.8)$$

and the mixed problem (M_1)

$$u''(t) + (a(t) + \lambda)\,u(t) = 0, \quad \text{a.e. } t \in [0, T], \quad u'(0) = u(T) = 0. \qquad (2.9)$$

Reasoning as in previous section it is clear that $u_1(\cdot, \lambda)$, the solution of

$$u''(t) + (a(t) + \lambda)\,u(t) = 0, \quad \text{a.e. } t \in [0, T],$$

satisfying that $u_1(0, \lambda) = 1$ and $u_1'(0, \lambda) = 0$ (see Definition 4) behaves in an analogous way to $u_2(\cdot, \lambda)$. Therefore, taking into account that the eigenvalues of (2.8) are those values of λ for which $u_1'(T, \lambda) = 0$, $u_1(\cdot, \lambda) \not\equiv 0$, and following the same argument than in the previous case, it can be deduced the existence of the following sequence of eigenvalues

$$\lambda_0^N[a] < \lambda_1^N[a] < \lambda_2^N[a] < \cdots < \lambda_k^N[a] < \cdots$$

for which problem (2.8), with $\lambda = \lambda_k^N[a]$, has a nontrivial solution $u_1(\cdot, \lambda_k^N[a])$ with exactly k zeros on $(0, T)$. Moreover, from Theorem 6, all the eigenvalues $\lambda_k^N[a]$ for $k = 0, 1, 2\ldots$ are simple.

The same way, the eigenvalues of (2.9) are those values of λ for which $u_1(T, \lambda) = 0$, $u_1(\cdot, \lambda) \not\equiv 0$, and it can be deduced the existence of the following sequence of eigenvalues

$$\lambda_0^{M_1}[a] < \lambda_1^{M_1}[a] < \lambda_2^{M_1}[a] < \cdots < \lambda_k^{M_1}[a] < \cdots$$

for which problem (2.9), with $\lambda = \lambda_k^{M_1}[a]$, has a nontrivial solution $u_1(\cdot, \lambda_k^{M_1}[a])$ with exactly k zeros on $(0, T)$. From Theorem 6 all the eigenvalues $\lambda_k^{M_1}[a]$ for $k = 0, 1, 2 \ldots$ are simple too.

Moreover, we could deduce the following alternating property

$$\lambda_0^N[a] < \lambda_0^{M_1}[a] < \cdots < \lambda_k^N[a] < \lambda_k^{M_1}[a] < \lambda_{k+1}^N[a] < \lambda_{k+1}^{M_1}[a] < \ldots$$

Finally, analogously to Lemma 2, we have the following results.

Lemma 3. *The function*

$$\lambda_n^{M_2} : L^\infty(I) \to \mathbb{R},$$

$$a \mapsto \lambda_n^{M_2}[a],$$

that assigns to every potential a the n-th eigenvalue, $n = 0, 1, \ldots$, of the corresponding mixed problem (M_2), is decreasing.

Lemma 4. *The function*

$$\lambda_n^{M_1} : L^\infty(I) \to \mathbb{R},$$

$$a \mapsto \lambda_n^{M_1}[a],$$

that assigns to every potential a the n-th eigenvalue, $n = 0, 1, \ldots$, of the corresponding mixed problem (M_1), is decreasing.

Lemma 5. *The function*

$$\lambda_n^N : L^\infty(I) \to \mathbb{R},$$

$$a \mapsto \lambda_n^N[a],$$

that assigns to every potential a the n-th eigenvalue, $n = 0, 1, \ldots$, of the corresponding Neumann problem, is decreasing.

2.5 SPECTRAL PROPERTIES OF THE PERIODIC PROBLEM: INTERVALS OF STABILITY AND INSTABILITY

This section is devoted to prove the following classical result, which is related to the spectrum of the Hill's equation.

Theorem 11 (Oscillation). *There exist two increasing sequences of real numbers*

$$\left\{\lambda_n^P[a]\right\}_{n=0}^{\infty} \quad and \quad \left\{\lambda_n^A[a]\right\}_{n=0}^{\infty}$$

such that Eq. (1.9) has a T-periodic solution if and only if $\lambda = \lambda_n^P[a]$, $n = 0, 1$, $2, \ldots$, and a T-antiperiodic solution if and only if $\lambda = \lambda_n^A[a]$, $n = 0, 1, 2, \ldots$.

Moreover, $\lambda_n^P[a]$, $n = 0, 1, \ldots$, are the roots of the equation $\Delta(\lambda) = 2$ and $\lambda_n^A[a]$, $n = 0, 1, \ldots$, those of $\Delta(\lambda) = -2$, where

$$\Delta(\lambda) = u_1(T, \lambda) + u_2'(T, \lambda),$$

being u_1 and u_2 the normalized solutions given in Definition 4.

The following inequalities hold

$$\lambda_0^P[a] < \lambda_0^A[a] \leq \lambda_1^A[a] < \lambda_1^P[a] \leq \lambda_2^P[a] < \lambda_2^A[a] \leq \lambda_3^A[a] < \lambda_3^P[a] \leq \lambda_4^P[a]\ldots$$

and, moreover,

$$\lim_{n\to\infty} \lambda_n^P[a] = \infty, \qquad \lim_{n\to\infty} \lambda_n^A[a] = \infty.$$

The trivial solution of (1.9) is stable if λ belongs to one of the following intervals

$$(\lambda_0^P[a], \lambda_0^A[a]), \ (\lambda_1^A[a], \lambda_1^P[a]), \ (\lambda_2^P[a], \lambda_2^A[a]), \ (\lambda_3^A[a], \lambda_3^P[a]), \ldots$$

On the other hand, if λ belongs to one of the intervals

$$(-\infty, \lambda_0^P[a]], \ (\lambda_0^A[a], \lambda_1^A[a]), \ (\lambda_1^P[a], \lambda_2^P[a]), \ (\lambda_2^A[a], \lambda_3^A[a]), \ldots$$

the trivial solution of (1.9) is unstable.

The trivial solution of (1.9) is stable for $\lambda = \lambda_{2k-1}^P[a]$ or $\lambda = \lambda_{2k}^P[a]$, $k = 1, 2, \ldots$, if and only if $\lambda_{2k-1}^P[a] = \lambda_{2k}^P[a]$. Analogously, it is stable for $\lambda = \lambda_{2k}^A[a]$ or $\lambda = \lambda_{2k+1}^A[a]$, $k = 0, 1, \ldots$, if and only if $\lambda_{2k}^A[a] = \lambda_{2k+1}^A[a]$.

Definition 8. The function

$$\Delta(\lambda) = u_1(T, \lambda) + u_2'(T, \lambda),$$

is known as the discriminant of the Hill's equation (1.9).

Definition 9. With the notation given in the previous theorem, the real numbers $\lambda_n^P[a]$ are known as the characteristic values of first type of (1.9) and the $\lambda_n^A[a]$ as characteristic values of second type.

Remark 9. The instability interval $(-\infty, \lambda_0^P[a]]$ always exists and it is called the zeroth instability interval. Any other instability interval could disappear, which happens when $\lambda_{2n+1}^P[a] = \lambda_{2n+2}^P[a]$ or $\lambda_{2n}^A[a] = \lambda_{2n+1}^A[a]$. This occurs, for example, when a is a constant function.

In order to prove the previous theorem, we will start by considering some properties of the discriminant $\Delta(\lambda)$.

Lemma 6. *All the roots of equations $\Delta(\lambda) = 2$ and $\Delta(\lambda) = -2$ are real and bigger than λ^*, with λ^* given in Theorem 4.*

Proof. First, we observe that $\Delta(\lambda) = 2$ is equivalent to $\rho_1 = \rho_2 = 1$ and $\Delta(\lambda) = -2$ is equivalent to $\rho_1 = \rho_2 = -1$. Indeed,

$$\Delta(\lambda) = u_1(T) + u_2'(T) = \pm 2 \Leftrightarrow \rho^2 \pm 2\rho + 1 = 0 \Leftrightarrow (\rho \pm 1)^2 = 0 \Leftrightarrow \rho = \pm 1.$$

As it is showed in the proof of Theorem 2, if $\Delta(\lambda) = \pm 2$, Eq. (1.9) will have a solution of the form $u(t) = e^{i\alpha t} p(t)$ (with p a T-periodic function, $\alpha = 0$ if $\rho_1 = 1$ and $\alpha = \frac{k\pi}{T}$, with k an odd integer, if $\rho_1 = -1$). In the proof of Theorem 3 it was seen that, when λ takes a complex value, there can not exist a solution $u(t) = e^{i\alpha t} p(t)$ with $\alpha \in \mathbb{R}$ and p a T-periodic function. So, we deduce that the roots of $\Delta(\lambda) = \pm 2$ must be real.

On the other hand, Lemma 1 assures that, when $\lambda \in \mathbb{R}$ and $\lambda \leq \lambda^*$, $\Delta(\lambda) > 2$. Consequently, all the roots of $\Delta(\lambda) = \pm 2$ are bigger than λ^*. $\square$

Lemma 7. *Eq. (1.9) has a T-periodic solution if and only if $\Delta(\lambda) = 2$ and a T-antiperiodic solution if and only if $\Delta(\lambda) = -2$.*

Proof. It is an immediate consequence of the fact that condition $\rho_1 = \rho_2 = 1$ given in Theorem 2 is equivalent to $\Delta(\lambda) = 2$, meanwhile $\rho_1 = \rho_2 = -1$ is equivalent to $\Delta(\lambda) = -2$. $\square$

Lemma 8. *Let μ be a root of equation $\Delta(\lambda) = 2$ such that $\Delta'(\mu) = \frac{d\Delta}{d\lambda}(\mu) \leq 0$. Then $\Delta'(\lambda) < 0$ on every interval (μ, μ_1^*) in which $\Delta(\lambda) > -2$.*

Analogously, let μ' be a root of $\Delta(\lambda) = -2$ such that $\Delta'(\mu') \geq 0$. Then $\Delta'(\lambda) > 0$ on every interval (μ', μ_1') in which $\Delta(\lambda) < 2$.

Proof. We divide the proof into several steps.

Step 1. We introduce the following notation

$$z_1(t, \lambda) = \frac{\partial}{\partial \lambda} u_1(t, \lambda),$$

$$z_1'(t, \lambda) = \frac{\partial}{\partial \lambda} u_1'(t, \lambda) = \frac{\partial}{\partial t} z_1(t, \lambda),$$

$$z_1''(t, \lambda) = \frac{\partial}{\partial \lambda} u_1''(t, \lambda) = \frac{\partial^2}{\partial t^2} z_1(t, \lambda),$$

$$z_2(t, \lambda) = \frac{\partial}{\partial \lambda} u_2(t, \lambda),$$

$$z_2'(t, \lambda) = \frac{\partial}{\partial \lambda} u_2'(t, \lambda) = \frac{\partial}{\partial t} z_2(t, \lambda),$$

$$z_2''(t, \lambda) = \frac{\partial}{\partial \lambda} u_2''(t, \lambda) = \frac{\partial^2}{\partial t^2} z_2(t, \lambda),$$

$$\eta_1(\lambda) = u_1(T, \lambda),$$

$$\eta_1'(\lambda) = u_1'(T, \lambda),$$

$$\eta_2(\lambda) = u_2(T, \lambda),$$

$$\eta_2'(\lambda) = u_2'(T, \lambda).$$

Moreover, we will write $\Delta(\lambda) = u_1(T, \lambda) + u_2'(T, \lambda) = \eta_1(\lambda) + \eta_2'(\lambda)$ and

$$\Delta'(\lambda) = \frac{\partial \Delta}{\partial \lambda}(\lambda) = z_1(T, \lambda) + z_2'(T, \lambda).$$

Step 2. We will prove the following equality

$$\Delta'(\lambda) = (\eta_1 - \eta_2') \int_0^T u_1(t) u_2(t) \, dt - \eta_2 \int_0^T u_1^2(t) \, dt + \eta_1' \int_0^T u_2^2(t) \, dt. \quad (2.10)$$

In order to do that, we will derive on Eq. (1.9) with respect to λ for $u = u_1$ and $u = u_2$, obtaining

$$z_1'' + (\lambda + a(t)) z_1 = -u_1, \quad (2.11)$$

$$z_2'' + (\lambda + a(t)) z_2 = -u_2. \quad (2.12)$$

Now, we have two nonhomogeneous linear differential equations of second order. To solve them, we note that the equation $z_1'' + (\lambda + a(t)) z_1 = -u_1$ is equivalent to the system

$$\begin{pmatrix} z_1' \\ z_1'' \end{pmatrix} = \begin{pmatrix} 0 & 1 \\ -(\lambda + a(t)) & 0 \end{pmatrix} \begin{pmatrix} z_1 \\ z_1' \end{pmatrix} + \begin{pmatrix} 0 \\ -u_1 \end{pmatrix}.$$

A particular solution for the previous system with $z_1(0) = z_1'(0) = 0$ can be found by using the method of variation of parameters, taking into account that the general solution of the associated homogeneous equation can be expressed in the form $c_1 u_1 + c_2 u_2$. So, a particular solution of the system would be the following:

$$\begin{pmatrix} z_1(t) \\ z_1'(t) \end{pmatrix} = M(t) \int_0^t M^{-1}(s) \begin{pmatrix} 0 \\ -u_1(s) \end{pmatrix} ds,$$

with

$$M(t) = \begin{pmatrix} u_1(t) & u_2(t) \\ u_1'(t) & u_2'(t) \end{pmatrix}$$

and

$$M^{-1}(t) = \frac{1}{W(u_1, u_2)(t)} \begin{pmatrix} u_2'(t) & -u_2(t) \\ -u_1'(t) & u_1(t) \end{pmatrix}$$

with the Wronskian $W(u_1, u_2)(t) = 1$ for all $t \in \mathbb{R}$.

Then, the solution of the system is given by the expression

$$z_1(t) = u_1(t) \int_0^t u_2(s) u_1(s) \, ds - u_2(t) \int_0^t u_1^2(s) \, ds,$$

$$z_1'(t) = u_1'(t) \int_0^t u_2(s) u_1(s) \, ds - u_2'(t) \int_0^t u_1^2(s) \, ds.$$

Analogously, the equation $z_2'' + (\lambda + a(t)) z_2 = -u_2$, $z_2(0) = z_2'(0) = 0$, has as a particular solution

$$z_2(t) = u_1(t) \int_0^t u_2^2(s) \, ds - u_2(t) \int_0^t u_1(s) u_2(s) \, ds,$$

$$z_2'(t) = u_1'(t) \int_0^t u_2^2(s) \, ds - u_2'(t) \int_0^t u_1(s) u_2(s) \, ds.$$

Equality (2.10) is immediately obtained from the previous ones by simply substituting $t = T$.

Step 3. We will find an equivalent expression of $\Delta'(\lambda)$.

Taking into account that the Wronskian $u_1(t) u_2'(t) - u_2(t) u_1'(t) = 1$ for all t and, substituting $t = T$, it is obtained that

$$\eta_1 \eta_2' - \eta_2 \eta_1' = 1,$$

so

$$\Delta^2 - 4 = (\eta_1 + \eta_2')^2 - 4(\eta_1 \eta_2' - \eta_2 \eta_1') = (\eta_1 - \eta_2')^2 + 4\eta_1' \eta_2. \qquad (2.13)$$

The function sign of η_1' is defined in the following way

$$\operatorname{sgn} \eta_1' = \begin{cases} 1 & \text{if} \quad \eta_1' > 0, \\ -1 & \text{if} \quad \eta_1' < 0, \\ 0 & \text{if} \quad \eta_1' = 0, \end{cases}$$

and, assuming that $\eta_1' \neq 0$, using expression (2.13), it is not difficult to verify that equality (2.10) can be rewritten in the following way

$$\Delta'(\lambda) = \operatorname{sgn} \eta_1' \left\{ \int_0^T \left(\sqrt{|\eta_1'|} \, u_2 + \operatorname{sgn} \eta_1' \frac{\eta_1 - \eta_2'}{2\sqrt{|\eta_1'|}} \, u_1 \right)^2 dt - \frac{\Delta^2 - 4}{4 |\eta_1'|} \int_0^T u_1^2 \, dt \right\}.$$

$$(2.14)$$

It is clear that Eq. (2.14) shows that $\Delta'(\lambda)$ and η_1' have the same sign on every interval in which $\eta_1' \neq 0$ and $\Delta^2 \leq 4$.

Step 4. Now we will consider μ such that $\Delta(\mu) = 2$ and $\Delta'(\mu) \leq 0$. We will prove that, for $\delta > 0$ small enough, $\Delta(\lambda)$ is decreasing for $\lambda \in (\mu, \mu + \delta)$.

This is obvious when $\Delta'(\mu) < 0$, so we will prove it in the case that $\Delta'(\mu) = 0$. Note that, under these conditions, the previous equality lets us conclude that $\eta_1'(\mu) = 0$ (because $\Delta^2(\mu) = 4$ and so $\eta_1'(\mu) \neq 0$ would imply that $\Delta'(\mu) \neq 0$). Taking into account that

$$0 = \Delta^2(\mu) - 4 = (\eta_1(\mu) - \eta_2'(\mu))^2 + 4\eta_1'(\mu)\,\eta_2(\mu) = (\eta_1(\mu) - \eta_2'(\mu))^2$$

and

$$2 = \Delta(\mu) = \eta_1(\mu) + \eta_2'(\mu),$$

it is deduced that

$$\eta_1(\mu) = \eta_2'(\mu) = 1.$$

Thus, (2.10) is reduced to

$$0 = \Delta'(\mu) = -\eta_2(\mu) \int_0^T u_1^2(t, \mu)\, dt$$

and, consequently,

$$\eta_2(\mu) = 0.$$

We will obtain now the value of $\Delta''(\lambda) = \frac{d\Delta'}{d\lambda}(\lambda)$ at $\lambda = \mu$.

To do that, we derive (2.10) with respect to λ and consider the expressions for z_1 and z_2, evaluating them at $t = T$. This way, we obtain the derivatives of η_1, η_2, η'_1, and η'_2 with respect to λ at $\lambda = \mu$, as follows

$$\frac{\partial \eta_1}{\partial \lambda}(\mu) = z_1(T, \mu) = u_1(T) \int_0^T u_2(s)\, u_1(s)\, ds - u_2(T) \int_0^T u_1^2(s)\, ds$$

$$= \int_0^T u_2(s)\, u_1(s)\, ds,$$

$$\frac{\partial \eta_2}{\partial \lambda}(\mu) = z_2(T, \mu) = u_1(T) \int_0^T u_2^2(s)\, ds - u_2(T) \int_0^T u_1(s)\, u_2(s)\, ds$$

$$= \int_0^T u_2^2(s)\, ds,$$

$$\frac{\partial \eta'_1}{\partial \lambda}(\mu) = z'_1(T, \mu) = u'_1(T) \int_0^T u_2(s)\, u_1(s)\, ds - u'_2(T) \int_0^T u_1^2(s)\, ds$$

$$= -\int_0^T u_1^2(s)\, ds,$$

$$\frac{\partial \eta'_2}{\partial \lambda}(\mu) = z'_2(T, \mu) = u'_1(T) \int_0^T u_2^2(s)\, ds - u'_2(T) \int_0^T u_1(s)\, u_2(s)\, ds$$

$$= -\int_0^T u_1(s)\, u_2(s)\, ds.$$

Therefore,

$$\Delta''(\mu) = 2 \left(\int_0^T u_1(s)\, u_2(s)\, ds \right)^2 - 2 \int_0^T u_1^2(s)\, ds \int_0^T u_2^2(s)\, ds.$$

Since u_1 and u_2 are linearly independent, Cauchy-Schwartz's inequality lets us conclude that $\Delta''(\mu) < 0$ and consequently Δ' is strictly decreasing on an interval $(\mu, \mu + \delta)$.

Conclusion. Suppose that the assertion is false. As $\Delta'(\mu) \leq 0$ and Δ' is decreasing on $(\mu, \mu + \delta)$, there will exist a minimum $\mu^* > \mu$ such that $\Delta'(\lambda) < 0$ for $\mu < \lambda < \mu^*$ and $\Delta'(\mu^*) = 0$ with $\Delta(\mu^*) > -2$.

In such a case, since $|\Delta(\mu^*)| < 2$,

$$0 > \Delta^2(\mu^*) - 4 = (\eta_1(\mu^*) - \eta'_2(\mu^*))^2 + 4\,\eta'_1(\mu^*)\,\eta_2(\mu^*),$$

from where we conclude that $\eta'_1(\mu^*)\,\eta_2(\mu^*) < 0$. Thus, $\eta'_1(\mu^*) \neq 0$ and, using equality (2.14), we conclude that $\Delta'(\mu^*) \neq 0$, which is a contradiction.

The behaviour of $\Delta(\mu) = -2$ is proved in an analogous way. $\square$

Corollary 3. *The roots of the equation*

$$\Delta^2(\lambda) - 4 = 0$$

can not have multiplicity bigger than two.

Proof. Let μ be a root of $\Delta^2(\lambda) - 4 = 0$ such that $\Delta'(\mu) = 0$. Then, the proof of Lemma 8, Step 4, establishes that $\Delta''(\mu) < 0$ when $\Delta(\mu) = 2$ and $\Delta''(\mu) > 0$ when $\Delta(\mu) = -2$. Thus, we conclude that μ can not have multiplicity bigger than two. $\square$

Lemma 9. *If $\lambda_0^P[a]$ is the smallest zero of the equation $\Delta^2(\lambda) - 4 = 0$, then $\lambda_0^P[a]$ is a simple root and $\Delta'(\lambda_0^P[a]) < 0$.*

Proof. If $\lambda_0^P[a]$ were not a simple root, then $\Delta'(\lambda_0^P[a]) = 0$. If this happened, the proof of Corollary 3 would allow us to affirm that $\Delta''(\lambda_0^P[a]) < 0$, so we would conclude that $\lambda_0^P[a]$ is a relative maximum for function Δ. This is a contradiction because, from Lemma 1, we know that $\Delta(\lambda) > 2$ for all $\lambda < \lambda_0^P[a]$.

Then necessarily $\Delta'(\lambda_0^P[a]) < 0$. $\square$

Now we are in a position to prove the Oscillation Theorem.

Proof of Theorem 11. We divide the proof into several steps.

Step 1. Functions $\Delta(\lambda) - 2$ and $\Delta(\lambda) + 2$ have an infinite number of zeros.

Consider equality (2.13) obtained in the proof of Lemma 8. This equality assures that if $\eta_1' = 0$, then $\Delta^2 \geq 4$. So, by assuming that $\eta_1' = 0$, taking into account that

$$1 = W(u_1, u_2) = \eta_1 \eta_2' - \eta_2 \eta_1' = \eta_1 \eta_2',$$

we deduce that $\eta_2' = \frac{1}{\eta_1}$. As a consequence, for all λ such that $\eta_1' = 0$, function Δ follows the expression

$$\Delta(\lambda) = \eta_1(\lambda) + \eta_2'(\lambda) = \frac{\eta_1^2(\lambda) + 1}{\eta_1(\lambda)},$$

in particular, $\Delta(\lambda)$ has the same sign as $\eta_1(\lambda) = u_1(T, \lambda)$.

We will see that there exist infinite values of λ for which $\eta_1' = u_1'(T, \lambda) = 0$.

$u_1(t, \lambda)$

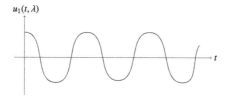

Figure 2.3 Graph of u_1 for λ large enough.

Reasoning in a similar way to Section 2.2, we can see that, for a fixed λ such that $a(t) + \lambda > 0$ for a.e. $t \in I$, u_1 oscillates and has an infinite number of zeros for $t > 0$, so it changes its sign every time it vanishes.

On the other hand, as u_1 solves (1.9) for λ large enough, such that $a(t) + \lambda > 0$ for a.e. $t \in I$, it satisfies that $u_1''(t) = -(a(t) + \lambda) u_1(t)$ for a.e. $t \in I$. Then, if $u_1(t) > 0$, the function will be concave in a neighbourhood of the point and if $u_1(t) < 0$, it will be convex. We deduce then that u_1' vanishes exactly once between each pair of consecutive zeros of u_1, that the points where u_1' vanishes are relative extrema of u_1 (since u_1'' has constant sign on a neighbourhood of those points because $u_1 \neq 0$) and that u_1' changes its sign every time it vanishes. That is, the solution $u_1(t, \lambda)$, considering a fixed λ large enough, oscillates in a similar way to Fig. 2.3.

If λ increases, the number of zeros of u_1 on the interval $[0, T]$ will increase too. This is a consequence of the Separation Theorem (Theorem 6), which assures that, for a fixed λ, u_1 vanishes exactly once between each pair of consecutive zeros of the solution of Dirichlet problem. Then, for $\lambda = \lambda_k^D[a]$, u_1 will have exactly $k + 1$ zeros on $[0, T]$.

Then, if λ grows up, the number of zeros of u_1' will also increase.

Consider now $u_{1_T}'(\lambda) \equiv u_1'(T, \lambda)$ as a function depending on λ. We will prove that it has a sequence of zeros which tends to infinity. To do that, suppose that there is $\bar{\lambda}$ such that $u_{1_T}' \neq 0$ for all $\lambda \geq \bar{\lambda}$. Then, let $\mu_1 \geq \bar{\lambda}$ and suppose that $u_{1_T}'(\mu_1) > 0$ and $u_1'(t, \mu_1)$ has k zeros on the interval $[0, T]$.

The existence of $\mu_2 > \mu_1 \geq \bar{\lambda}$ such that $u_1'(t, \mu_2)$ has $k + 1$ zeros on the interval $[0, T]$ is assured. From the remarks made about the way in which u_1 oscillates, necessarily $u_{1_T}'(\mu_2) < 0$. As $u_1'(t, \lambda)$ is continuous and, consequently, $u_{1_T}'(\lambda)$ is continuous too, we can apply Bolzano's Theorem and affirm that there exists $\mu_0 \in (\mu_1, \mu_2)$ such that $u_{1_T}'(\mu_0) = 0$.

Then we deduce that u_{1_T}' vanishes at infinite $\lambda's$ which extend through the positive real axis. Moreover, from what we have just seen, it can be deduced that $u_1(T, \lambda)$ changes its sign exactly once between two consecutive zeros of u'_{1_T} for λ large enough. Therefore, there exist infinite values of λ, as big as we want, for which $u'_1(T, \lambda) = 0$, $u_1(T, \lambda) > 0$ and, consequently, $\Delta(\lambda) \geq 2$. The same way, there exists an infinite number of values of λ, as big as we want, for which $\Delta(\lambda) \leq -2$.

As a consequence, we conclude that functions $\Delta(\lambda) + 2$ and $\Delta(\lambda) - 2$ have infinite zeros $\lambda_n^P[a]$ and $\lambda_n^A[a]$, respectively, as large as we want. Therefore, there exist two sequences of real numbers,

$$\left\{\lambda_n^P[a]\right\}_{n=0}^{\infty} \quad \text{and} \quad \left\{\lambda_n^A[a]\right\}_{n=0}^{\infty},$$

satisfying that

$$\lim_{n \to \infty} \lambda_n^P[a] = \infty \quad \text{and} \quad \lim_{n \to \infty} \lambda_n^A[a] = \infty.$$

Step 2. The following inequalities hold

$$\lambda_0^P[a] < \lambda_0^A[a] \leq \lambda_1^A[a] < \lambda_1^P[a] \leq \lambda_2^P[a] < \lambda_2^A[a] \leq \lambda_3^A[a] < \lambda_3^P[a] \leq \lambda_4^P[a] < \ldots$$
$$(2.15)$$

As showed in Lemma 1, if $\lambda \leq \lambda^*$, then $\Delta(\lambda) > 2$. So, the infinite set of zeros of function $\Delta(\lambda) - 2$ has a minimum, which we have denoted by $\lambda_0^P[a]$ and verifies that $\Delta'(\lambda_0^P[a]) < 0$ (Lemma 9). Consequently, Lemma 8 proves that function Δ decreases for $\lambda > \lambda_0^P[a]$ until it reaches a point $\lambda_0^A[a] > \lambda_0^P[a]$ such that $\Delta(\lambda_0^A[a]) = -2$. The existence of such $\lambda_0^A[a]$ is assured because function $\Delta(\lambda) = -2$ has a sequence of zeros unbounded from above. At this point, there exist two possibilities:

- If $\Delta'(\lambda_0^A[a]) = 0$, then $\lambda_0^A[a]$ is a double root of $\Delta(\lambda) + 2$ and it will occur that $\lambda_1^A[a] = \lambda_0^A[a]$ (as we have showed that $\Delta(\lambda) + 2$ can not have any root of multiplicity bigger than two).
- If $\Delta'(\lambda_0^A[a]) < 0$, then $\Delta(\lambda) < -2$ on the interval $(\lambda_0^A[a], \lambda_1^A[a])$, where $\lambda_1^A[a]$ is the smallest zero of $\Delta(\lambda) + 2$ which is bigger than $\lambda_0^A[a]$.

Now, we have that $\Delta'(\lambda_1^A[a]) \geq 0$, which can be deduced from $\Delta(\lambda) < \Delta(\lambda_1^A[a])$ for $\lambda < \lambda_1^A[a]$ close enough to $\lambda_1^A[a]$. As we have proved in Lemma 8, under these conditions we can affirm that Δ is increasing on any interval $(\lambda_1^A[a], \lambda_*)$ in which $\Delta(\lambda) < 2$. The maximal interval satisfying this condition is $(\lambda_1^A[a], \lambda_1^P[a])$, with $\lambda_1^P[a]$ the smallest root of $\Delta(\lambda) - 2$ which is bigger than $\lambda_1^A[a]$.

We could continue with this process indefinitely, which lets us conclude the veracity of (2.15).

Step 3. Functions $\Delta(\lambda) - 2$ and $\Delta(\lambda) + 2$ can not have an infinite number of zeros on a bounded interval.

Suppose that $\Delta(\lambda) - 2$ has an infinite sequence of zeros $\{\lambda_{n_k}\}_{k=0}^{\infty}$ which is bounded from above, that is, there exists $\tilde{\lambda}$ such that $\lambda_{n_k} \xrightarrow[k \to \infty]{} \tilde{\lambda}$. By continuity, it is clear that $\Delta(\tilde{\lambda}) = 2$.

On the other hand, (2.15) assures the existence of an infinite sequence of zeros of $\Delta(\lambda) + 2$, $\{\lambda_{m_k}^A\}_{k=0}^{\infty}$, which alternate with $\{\lambda_{n_k}^P\}_{k=0}^{\infty}$. As a consequence, $\lambda_{m_k}^A \xrightarrow[k \to \infty]{} \tilde{\lambda}$ and so $\Delta(\tilde{\lambda}) = -2$, which is a contradiction.

Step 4. Existence of stability intervals.

Corollary 2 establishes all the possible situations in which the trivial solution of Eq. (1.9) is stable. First condition in such corollary is satisfied if and only if λ is such that $|\Delta(\lambda)| < 2$. As an immediate consequence we deduce that the trivial solution of (1.9) is stable if λ belongs to one of the following intervals

$$(\lambda_0^P[a], \lambda_0^A[a]), (\lambda_1^A[a], \lambda_1^P[a]), (\lambda_2^P[a], \lambda_2^A[a]), (\lambda_3^A[a], \lambda_3^P[a]), \dots$$

Second condition in Corollary 2 is satisfied if and only if $\Delta(\lambda) = \pm 2$ and $\Delta'(\lambda) = 0$. This can be immediately deduced from Eqs. (2.10) and (2.13), taking into account the results obtained in Step 4 in the proof of Lemma 8. In consequence, the trivial solution of (1.9) is stable for $\lambda = \lambda_{2k+1}^P[a]$ ($\lambda = \lambda_{2k}^A[a]$), $k = 0, 1, \dots$, if and only if $\lambda_{2k+1}^P[a] = \lambda_{2k+2}^P[a]$ ($\lambda_{2k}^A[a] = \lambda_{2k+1}^A[a]$).

Finally, we deduce that the trivial solution of (1.9) is unstable if λ belongs to the intervals

$$(-\infty, \lambda_0^P[a]], (\lambda_0^A[a], \lambda_1^A[a]), (\lambda_1^P[a], \lambda_2^P[a]), (\lambda_2^A[a], \lambda_3^A[a]), \dots$$

On the other hand, the existence of T-periodic and T-antiperiodic solutions if and only if $\lambda = \lambda_n^P[a]$ or $\lambda = \lambda_n^A[a]$, respectively, has already been proved in Lemma 7.

Graphically, $\Delta(\lambda)$ would have an appearance similar to Fig. 2.4. □

Corollary 4. *If equation $\Delta^2(\lambda) - 4 = 0$ has a double root at $\lambda = \bar{\lambda}$, then either Hill's equation for $\lambda = \bar{\lambda}$ has two linearly independent T-periodic solutions (when $\Delta(\bar{\lambda}) = 2$) or two linearly independent T-antiperiodic solutions (when $\Delta(\bar{\lambda}) = -2$).*

Proof. We have just showed in Step 3 in the previous proof that $\Delta^2(\lambda) - 4 = 0$ has a double root if and only if

$$u_1(T, \lambda) + u_2'(T, \lambda) = \pm 2, \quad u_2(T, \lambda) = 0 \quad \text{and} \quad u_1'(T, \lambda) = 0.$$

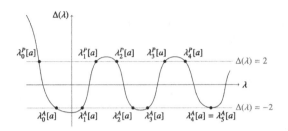

Figure 2.4 Graphic of $\Delta(\lambda)$.

Thus, we are in the case $\rho_1 = \rho_2$ and $\theta = 0$ in Theorem 2. This theorem assures that Hill's equation has a solution which is T-periodic (if $\rho_1 = \rho_2 = 1$) or T-antiperiodic (if $\rho_1 = \rho_2 = -1$). There exists also a linearly independent solution u satisfying that $u(t + T) = \rho_1 u(t)$. Thus u is also T-periodic if $\rho_1 = 1$ and T-antiperiodic if $\rho_1 = -1$.   □

2.6 RELATION BETWEEN EIGENVALUES OF NEUMANN, DIRICHLET, PERIODIC, AND ANTIPERIODIC PROBLEMS

Finally, using the characterization of the periodic and antiperiodic eigenvalues in terms of the value of the discriminant $\Delta(\lambda)$, it is possible to prove some alternating relation between the eigenvalues of these two problems with those of Neumann and Dirichlet.

As it was proved in Section 2.4, $\lambda_k^N[a]$ is an eigenvalue of Neumann problem (2.8) if and only if $u_1'(T, \lambda_k^N[a]) = 0$, $u_1(\cdot, \lambda_k^N[a]) \not\equiv 0$. In this situation the Wronskian reduces to

$$
\begin{aligned}
W(u_1, u_2)(T) &= u_1(T, \lambda_k^N[a])\, u_2'(T, \lambda_k^N[a]) - u_1'(T, \lambda_k^N[a])\, u_2(T, \lambda_k^N[a]) \\
&= u_1(T, \lambda_k^N[a])\, u_2'(T, \lambda_k^N[a]),
\end{aligned}
$$

and since $W(u_1, u_2) \equiv 1$, we deduce that

$$
u_2'(T, \lambda_k^N[a]) = \frac{1}{u_1(T, \lambda_k^N[a])}
$$

so the discriminant could be rewritten as

$$
\Delta(\lambda_k^N[a]) = u_1(T, \lambda_k^N[a]) + \frac{1}{u_1(T, \lambda_k^N[a])}.
$$

Since $x + \frac{1}{x} \geq 2$ for all $x > 0$, it is clear that $\left| \Delta(\lambda_k^N[a]) \right| \geq 2$. Moreover, since $u_1(\cdot, \lambda_k^N[a])$ has exactly k zeros on $(0, T)$ and $u_1(0, \lambda_k^N[a]) > 0$, it is easy to deduce that $u_1(T, \lambda_k^N[a]) > 0$ if k is even and $u_1(T, \lambda_k^N[a]) < 0$ if k is odd. Therefore

$$\Delta(\lambda_k^N[a]) \begin{cases} \geq 2, & \text{if } k \text{ is even} \\ \leq -2, & \text{if } k \text{ is odd} \end{cases}.$$

Remark 10. From the previous expression it is immediately deduced that between two consecutive eigenvalues of Neumann problem there must exist at least one eigenvalue of the periodic problem and one of the antiperiodic problem.

Analogously, since $\lambda_k^D[a]$ is an eigenvalue of Dirichlet problem if and only if $u_2(T, \lambda_k^D[a]) = 0$, $u_2(\cdot, \lambda_k^D[a]) \not\equiv 0$, in this situation we could also rewrite the discriminant as

$$\Delta(\lambda_k^D[a]) = u_1(T, \lambda_k^D[a]) + \frac{1}{u_1(T, \lambda_k^D[a])}.$$

In this case, $u_2(\cdot, \lambda_k^D[a])$ has k zeros on $(0, T)$ and vanishes at the boundary points. Consequently, from Sturm's Separation Theorem, $u_1(\cdot, \lambda_k^D[a])$ has exactly $k + 1$ zeros on $(0, T)$ and we deduce that

$$\Delta(\lambda_k^D[a]) \begin{cases} \geq 2, & \text{if } k \text{ is odd} \\ \leq -2, & \text{if } k \text{ is even} \end{cases}.$$

Remark 11. From the previous comments, it could be deduced that between two consecutive eigenvalues of Dirichlet problem there must exist at least one eigenvalue of the periodic problem and one of the antiperiodic problem.

It could also be proved (as a particular case of Theorems 13.7 and 13.8 in [6]) that

Theorem 12. *If u_k^P denotes the eigenfunction corresponding to the eigenvalue $\lambda_k^P[a]$ of the periodic problem, then*
1. *The eigenfunction u_0^P has no zeros on $[0, T]$;*
2. *Both u_{2n-1}^P and u_{2n}^P have exactly $2n$ zeros on $[0, T)$ for $n \geq 1$.*

Theorem 13. *If u_k^A denotes the eigenfunction corresponding to the eigenvalue $\lambda_k^A[a]$ of the antiperiodic problem, then both u_{2n}^A and u_{2n+1}^A have exactly $2n + 1$ zeros on $[0, T)$ for $n \geq 0$.*

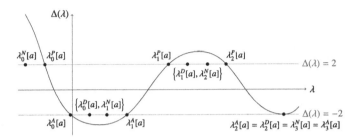

Figure 2.5 Periodic, Antiperiodic, Neumann, and Dirichlet eigenvalues.

Taking into account the restriction on the number of zeros of eigenfunctions of the periodic and antiperiodic problem and Remarks 10 and 11, through standard arguments using the Sturm's Comparison Theorem, Theorem 10 (analogously to Sections 2.3 and 2.4), it could be proved the following alternating order of eigenvalues

$$\lambda_0^N[a] \leq \lambda_0^P[a] < \lambda_0^A[a] \leq \begin{Bmatrix} \lambda_1^N[a] \\ \lambda_0^D[a] \end{Bmatrix} \leq \cdots$$

$$\cdots < \lambda_k^A[a] \leq \begin{Bmatrix} \lambda_{k+1}^N[a] \\ \lambda_k^D[a] \end{Bmatrix} \leq \lambda_{k+1}^A[a] < \lambda_{k+1}^P[a] \leq \begin{Bmatrix} \lambda_{k+2}^N[a] \\ \lambda_{k+1}^D[a] \end{Bmatrix}$$

$$\leq \lambda_{k+2}^P[a] < \lambda_{k+2}^A[a].$$

This situation is represented in Fig. 2.5.

Remark 12. Note that a periodic (or antiperiodic) eigenvalue has multiplicity two if and only if it is a periodic (or antiperiodic), Dirichlet, and Neumann eigenvalue at the same time. In this case, this periodic (antiperiodic) eigenvalue will have two independent eigenfunctions, one of them will be the eigenfunction of Neumann problem and the other one will be the eigenfunction of Dirichlet problem.

Remark 13. Along this chapter we have assumed that the potential is the T-periodic extension of $a \in L^\infty(I)$. This assumption allows us to ensure the existence of $\lambda_1 < \lambda_2$ such that $a(t) + \lambda_1 < 0 < a(t) + \lambda_2$ for a.e. $t \in I$. Such property is used to ensure that both the Dirichlet and the periodic problems have a divergent sequence of real eigenvalues, bounded from below.

If we consider the potential a as the T-periodic extension of an $L^\alpha(I)$-function, $\alpha \geq 1$, the existence of such λ_1 and λ_2 cannot be ensured in general. In this case the sequence of eigenvalues can be deduced from the construction of a related integral operator characterized by a continuous kernel, which is known as its Green's function (see [2] for details). By means of classical theory of functional analysis [1,7], one can verify that such integral operator is compact and self-adjoint in $L^2(I)$. So, as a consequence of [1, Theorem 6.11], we can ensure the existence of an increasing and divergent sequence of eigenvalues for both problems. The fact that there is a nontrivial eigenfunction, strictly positive on $(0, T)$, related to the smallest eigenvalue, follows from the Krein-Rutman Theorem [1,7].

Concerning the Dirichlet problem, the fact that the eigenfunction u_k related to the eigenvalue $\lambda_k^D[a]$ has exactly k zeros on $(0, T)$, is a direct consequence of Sturm's separation and comparison (Theorems 6 and 10). From this fact, in an analogous way to Section 2.4, it can be deduced the existence of the sequence of eigenvalues and eigenfunctions, with their corresponding number of zeros, for Neumann and mixed Problems. As a consequence, the decreasing character of the eigenvalues for Dirichlet, Neumann, and mixed problems, showed in Lemmas 2, 3, 4, and 5, remains valid for $a \in L^\alpha(I)$.

It is important to note that the existence of this sequence of eigenvalues, with their correspondent eigenfunctions, remains valid for a more general self-adjoint problem. In fact, in [8, Chapter 4], it is proved a general result that includes the following Sturm-Liouville equation:

$$-(p\,u')'(t) + q(t)\,u(t) = \lambda\,w(t)\,u(t), \quad \text{a.e. } t \in (c, d), \quad -\infty \leq c < d \leq \infty,$$

$$(2.16)$$

with p, q, and w real valued functions such that $1/p$, q, $w \in L^1(c, d)$ and $w > 0$ a.e. on (c, d).

Moreover, the following two types of boundary conditions are considered:

$$u(d) = k_{11}\,u(c) + k_{12}\,(p\,u')(c), \quad (p\,u')(d) = k_{21}\,u(c) + k_{22}\,(p\,u')(c), \quad (2.17)$$

with

$$k_{11}\,k_{22} - k_{21}\,k_{12} = 1, \qquad (2.18)$$

which are denoted as Coupled Self-adjoint boundary conditions, and the Separated self-adjoint boundary conditions:

$$A_1\,u(c) + A_2\,(p\,u')(c) = 0, \quad B_1\,u(d) + B_2\,(p\,u')(d) = 0, \qquad (2.19)$$

with A_1, A_2, B_1, $B_2 \in \mathbb{R}$ satisfying $(A_1, A_2) \neq (0, 0)$ and $(B_1, B_2) \neq (0, 0)$.

Thus, in [8, Theorem 4.3.1] the following results are proved:

1. All the eigenvalues of problem (2.16) related to boundary conditions coupled (2.17) or separated (2.19), are real, isolated with no finite accumulation point, and there is an infinite but countable number of them.

2. If $p > 0$ on (c, d) and the periodic boundary conditions (2.17) are fulfilled, then the eigenvalues are bounded from below and can be ordered to satisfy

$$-\infty < \lambda_0 \leq \lambda_1 \leq \lambda_2 \leq \cdots; \qquad \lambda_k \to \infty, \text{ as } k \to \infty. \qquad (2.20)$$

Each eigenvalue may be simple or double but there cannot be two consecutive equalities in (2.20) since, for any value of λ, Eq. (2.16) has exactly two linearly independent solutions. Note that λ_k is well defined for each $k \geq 0$ but there is some arbitrariness in the indexing of the eigenfunctions corresponding to a double eigenvalue since every nontrivial solution of the equation for such an eigenvalue is an eigenfunction. Given such an indexing scheme, let u_k be a real-valued eigenfunction of λ_k for the coupled conditions (2.18), $k \geq 0$, then the number of zeros of u_k in (c, d) is 0 or 1, if $k = 0$, and $k - 1$ or k or $k + 1$ if $k \geq 1$.

3. If $p > 0$ and the boundary conditions are the separated ones (2.19) then strict inequality holds everywhere in (2.20). Furthermore, if u_k is an eigenfunction of λ_k, then u_k is unique up to constant multiples and has exactly k zeros in the open interval (c, d).

It is important to point out that the coupled conditions (2.17) cover the periodic boundary conditions ($k_{11} = k_{22} = 1$, $k_{21} = k_{12} = 0$). In this case, if $c, d \in \mathbb{R}$, the Krein-Rutman Theorem [1,7] ensures that the least eigenvalue is simple with its corresponding eigenfunction strictly positive on (c, d) and that the rest of the eigenfunctions change its sign on (c, d).

Note also that coupled conditions (2.17) cover also the antiperiodic boundary conditions ($k_{11} = k_{22} = -1$, $k_{21} = k_{12} = 0$). In this case, the Krein-Rutman Theorem is not applicable (because the corresponding Green's function always changes its sign) and, as we have seen in this chapter, the first eigenvalue can be not simple and the corresponding eigenfunctions can change its sign on the interval of definition.

We want to point out that we have made the proofs for a T-periodic extension of $a \in L^{\infty}(I)$ because it is based on the elementary oscillation theory and it does not need to apply more sophisticated tools in functional analysis and spectral theory.

REFERENCES

[1] H. Brézis, Functional Analysis, Sobolev Spaces and Partial Differential Equations, Universitext, Springer, New York, 2011.

[2] A. Cabada, Green's Functions in the Theory of Ordinary Differential Equations, Springer Briefs Math., 2014.

[3] J.K. Hale, Ordinary Differential Equations, second edition, Robert E. Krieger Publishing Co., New York, 1980.

[4] W. Magnus, S. Winkler, Hill's Equation, Dover Publications, Inc., New York, 1979.

[5] G.F. Simmons, J.S. Robertson, Differential Equations With Applications and Historical Notes, McGraw-Hill, 1991.

[6] J. Weidmann, Spectral Theory of Ordinary Differential Operators, Springer-Verlag, Berlin, Heidelberg, 1987.

[7] E. Zeidler, Nonlinear Functional Analysis and Its Applications. I. Fixed-Point Theorems, Springer-Verlag, New York, 1986.

[8] A. Zettl, Sturm-Liouville Theory, Math. Surveys Monogr., vol. 121, American Mathematical Society, Providence, RI, 2005.

CHAPTER 3

Nonhomogeneous Equation

Contents

3.1	Introduction	49
3.2	The Green's Function	52
3.3	Periodic Conditions	58
	3.3.1 Properties of the Periodic Green's Function	58
	3.3.2 Optimal Conditions for the Periodic MP and AMP	63
	3.3.3 Explicit Criteria for the Periodic AMP and MP	68
	3.3.4 More on Explicit Criteria	72
	3.3.5 Examples	75
3.4	Non-Periodic Conditions	80
	3.4.1 Neumann Problem	81
	3.4.2 Dirichlet Problem	84
	3.4.3 Relation Between Neumann and Dirichlet Problems	85
	3.4.4 Mixed Problems and their Relation with Neumann and Dirichlet Ones	86
	3.4.5 Order of Eigenvalues and Constant Sign of the Green's Function	91
	3.4.6 Relations Between Green's Functions. Comparison Principles	100
	3.4.7 Constant Sign for Non-Periodic Green's Functions	103
	3.4.8 Global Order of Eigenvalues	107
	3.4.9 Examples	109
3.5	General Second Order Equation	113
	3.5.1 Periodic Problem	113
	3.5.2 Non-Periodic Conditions	120
References		127

3.1 INTRODUCTION

The topic of maximum and antimaximum principles related to the Hill's operator

$$L[a]u(t) \equiv u''(t) + a(t)\,u(t), \quad t \in [0, T] \equiv I,$$

with

$$a : \mathbb{R} \to \mathbb{R}, \quad a \in L^{\alpha}(I),\, \alpha \geq 1 \quad \text{and} \quad a(t+T) = a(t) \quad \text{a.e. } t \in \mathbb{R}, \quad (3.1)$$

has been widely studied in the literature [1,5–7,17,26,30,36,37]. These comparison principles are fundamental tools when we consider nonlinear boundary value problems and apply, among others, monotone iterative techniques [18,21,31], lower and upper solutions method [3,14] or fixed points theorems [18,30], as we will see in Chapter 4. Moreover, comparison principles are fundamental in the study of partial differential equations, see [32], and are also relevant to stability theory, see [37].

To fix ideas, we will denote by $W^{2,1}(I)$ the space of functions $u \in \mathcal{C}^1(I)$ whose first derivative is an absolutely continuous function on I and let X be the Banach space

$$X = \left\{ u \in W^{2,1}(I) : U_i(u) \equiv \sum_{j=0}^{1} \left(\alpha_j^i u^{(j)}(0) + \beta_j^i u^{(j)}(T) \right) = 0, \quad i = 1, 2 \right\},$$

being α_j^i, β_j^i real constants for $i = 1, 2$ and $j = 0, 1$.

We say that $L[a]$ admits the maximum principle (MP) in X if and only if

$$u \in X, \ L[a]u \geq 0 \text{ on } I \Longrightarrow u \equiv 0 \quad \text{or} \quad u < 0 \quad \text{on } (0, T),$$

and $L[a]$ admits the antimaximum principle (AMP) in X if and only if

$$u \in X, \ L[a]u \geq 0 \text{ on } I \Longrightarrow u \equiv 0 \quad \text{or} \quad u > 0 \quad \text{on } (0, T).$$

Remark 14. Previous definitions are formulated in terms of the general Banach space X, which includes several kinds of boundary conditions. However, with some of the boundary conditions considered in this book, the strict constant sign could be extended to 0 and (or) T. In particular, if we consider

$$X_P = \left\{ u \in W^{2,1}(I) : u(0) = u(T), \ u'(0) = u'(T) \right\},$$

we have that $L[a]$ admits the maximum principle in X_P if and only if

$$u \in X_P, \ L[a]u \geq 0 \text{ on } I \Longrightarrow u \equiv 0 \quad \text{or} \quad u < 0 \quad \text{on } I,$$

and $L[a]$ admits the antimaximum principle in X_P if and only if

$$u \in X_P, \ L[a]u \geq 0 \text{ on } I \Longrightarrow u \equiv 0 \quad \text{or} \quad u > 0 \quad \text{on } I.$$

The same situation occurs when considering

$$X_N = \left\{ u \in W^{2,1}(I) : u'(0) = u'(T) = 0 \right\}.$$

Analogously, with

$$X_{M_1} = \left\{ u \in W^{2,1}(I) : u'(0) = u(T) = 0 \right\}$$

and

$$X_{M_2} = \left\{ u \in W^{2,1}(I) : \ u(0) = u'(T) = 0 \right\},$$

the strict constant sign is extended to $[0, T)$ and $(0, T]$, respectively.

We say that *operator $L[a]$ is nonresonant in X* if and only if the homogeneous problem

$$L[a]\, u(t) = 0, \quad t \in I, \quad u \in X, \tag{3.2}$$

has only the trivial solution. Of course, if $L[a]$ satisfies MP or AMP then it is nonresonant.

Some notation is needed throughout the chapter:

- We will denote by $h \succ 0$ a function $h \in L^{\alpha}(I)$ such that $h(t) \geq 0$ for a.e. $t \in I$ and $h \not\equiv 0$ on I.
- The positive part

$$h_+(t) = \max\{h(t), 0\}, \quad t \in I$$

and the negative part

$$h_-(t) = -\min\{h(t), 0\}, \quad t \in I$$

are defined as usual.

- For $1 < \alpha < \infty$ we denote by α^* its conjugate, that is, $\dfrac{1}{\alpha} + \dfrac{1}{\alpha^*} = 1$. If $\alpha = 1$ then $\alpha^* = \infty$ and vice-versa.
- Finally, denoting $H_0^1(I)$ as the usual Sobolev space of the $W^{1,2}(I)$ functions that satisfy the Dirichlet conditions, we define $K(\alpha, T)$ as the best Sobolev constant in the inequality

$$C\,\|u\|_{\alpha}^2 \leq \|u'\|_2^2 \quad \text{for all } u \in H_0^1(I),$$

which is given explicitly by (see Appendix A)

$$K(\alpha, T) = \begin{cases} \dfrac{2\pi}{\alpha\, T^{1+\frac{2}{\alpha}}} \left(\dfrac{2}{2+\alpha}\right)^{1-\frac{2}{\alpha}} \left(\dfrac{\Gamma\left(\frac{1}{\alpha}\right)}{\Gamma\left(\frac{1}{2}+\frac{1}{\alpha}\right)}\right)^2, & 1 \leq \alpha < \infty, \\[4mm] \dfrac{4}{T}, & \alpha = \infty. \end{cases} \tag{3.3}$$

We note that $H_0^1(I)$ is a Hilbert space.

3.2 THE GREEN'S FUNCTION

It is well-known that if $L[a]$ is nonresonant in X then for all $\sigma \in L^1(I)$ the problem

$$L[a]u(t) = \sigma(t), \quad t \in I, \quad u \in X, \tag{3.4}$$

has a unique solution $u \in W^{2,1}(I)$ and there exists a unique continuous function

$$G[a] : I \times I \to \mathbb{R},$$

such that

$$u(t) = \int_0^T G[a](t, s)\,\sigma(s)\,ds, \qquad \forall\, t \in I. \tag{3.5}$$

This function $G[a]$ is the so-called Green's function related to the operator $L[a]$ in X.

Moreover, it is also very well-known ([4,13]) that operator $L[a]$ is non-resonant and self-adjoint in X if and only if its related Green's function exists and is symmetrical with respect to the diagonal of its square of definition, that is,

$$G[a](t, s) = G[a](s, t), \quad \forall\, (t, s) \in I \times I.$$

We can think the Green's function $G[a](t, s)$ as the response of (3.4) to δ_s, the unit Dirac measure located at s, that is,

$$G[a](t, s) = L[a]^{-1}(\delta_s)(t).$$

Let us see an example: consider the boundary value problem

$$u'' = \sigma(t), \quad t \in [0, 1], \quad u(0) = u(1) = 0.$$

Then, the Green's function $G[0](t, s)$ is the solution of the problem

$$u'' = \delta_s(t), \quad t \in [0, 1], \quad u(0) = u(1) = 0,$$

and by direct integration we obtain that

$$u'(t) = \begin{cases} c(s), & t < s, \\ c(s) + 1, & t > s, \end{cases}$$

and then

$$u(t) = \begin{cases} c(s)\,t, & t < s, \\ t - s + c(s)\,t, & t > s, \end{cases}$$

where we have used the continuity of u on $[0, 1]$ and that $u(0) = 0$.

Now, taking into account that $u(1) = 0$ we get $c(s) = s - 1$. Therefore,

$$G[0](t, s) = \begin{cases} (s - 1)\,t, & t < s, \\ (t - 1)\,s, & t > s. \end{cases}$$

Heuristically, since $\sigma(t) = \int_0^1 \delta_s(t)\,\sigma(s)\,ds$ can be interpreted as a continuous superposition of an infinite number of impulses concentrated at s, we would expect, assuming that the superposition principle is valid in this framework, that the solution of (3.4) is given by the expression

$$u(t) = \int_0^T G[0](t, s)\,\sigma(s)\,ds, \qquad \forall\, t \in [0, 1].$$

That is, the solution of the nonhomogeneous problem (3.4) is given by an integral equation whose kernel is the Green's function. Moreover, by construction, $G[0](\cdot, s)$ satisfies the homogeneous differential equation except at $t = s$,

$$\frac{\partial^2 G[0]}{\partial t^2}(t, s) = 0, \qquad \forall t \neq s,$$

coupled with the boundary conditions,

$$G[0](0, s) = G[0](1, s) = 0.$$

Moreover, $G[0](t, s)$ is continuous and $\frac{\partial G[0]}{\partial t}(t, s)$ has a unitary jump discontinuity at the impulse point $t = s$.

These properties, obtained in a heuristic way for the preceding example, turn in fact to be the key properties that characterize the Green's function.

Definition 10. We say that $G[a]$ is a *Green's function* for problem (3.2) if it satisfies the following properties:

(G1) $G[a]$ is continuous on the square $I \times I$.

(G2) Both $\dfrac{\partial\, G[a]}{\partial t}$ and $\dfrac{\partial^2 G[a]}{\partial t^2}$ exist and are continuous on the triangles $0 \leq s < t \leq T$ and $0 \leq t < s \leq T$.

(G3) For each $t \in (0, T)$ there exist the lateral limits

$$\frac{\partial G[a]}{\partial t}(t, t^+) \quad \text{and} \quad \frac{\partial G[a]}{\partial t}(t, t^-)$$

(i.e., the limits when $(t, s) \to (t, t)$ with $s > t$ and with $s < t$). Moreover

$$\frac{\partial G[a]}{\partial t}(t, t^+) - \frac{\partial G[a]}{\partial t}(t, t^-) = -1$$

or, which is equivalent,

$$\frac{\partial G[a]}{\partial t}(t^+, t) - \frac{\partial G[a]}{\partial t}(t^-, t) = 1.$$

(G4) For each $s \in (0, T)$, the function $t \to G[a](t, s)$ is a solution of the differential equation $L[a]u = 0$ on $t \in [0, s)$ and $t \in (s, T]$. That is,

$$\frac{\partial^2 G[a]}{\partial t^2}(t, s) + a(t)\, G[a](t, s) = 0,$$

on both intervals.

(G5) For each $s \in (0, T)$, the function $t \to G[a](t, s)$ satisfies the boundary conditions $U_i(G[a](\cdot, s)) = 0$, $i = 1, 2$, that is,

$$\sum_{j=0}^{1} \left(\alpha_j^i \frac{\partial^j G[a]}{\partial t^j}(0, s) + \beta_j^i \frac{\partial^j G[a]}{\partial t^j}(T, s) \right) = 0, \qquad i = 1, 2.$$

The proof of the following theorem, confirming our heuristic approach, can be found in [8]. A general and detailed analysis of Green's functions related to ordinary differential equations has been done in [4]. For the role of the Green's function in partial differential equations see for instance [25].

Theorem 14. *Let us suppose that the homogeneous problem (3.2) has only the trivial solution. Then there exists a unique Green's function, G[a], related to (3.2).*

Moreover, for each function $\sigma \in L^1(I)$, the unique solution of problem (3.4) is given by the expression

$$u(t) = \int_0^T G[a](t, s)\, \sigma(s)\, ds, \qquad t \in [0, T].$$

Next, we prove a necessary condition that must be satisfied by the Green's function of a self-adjoint operator. This result can be found in [7].

Proposition 2. *Assume that operator $L[a]$ is nonresonant and self-adjoint on X. If the Green's function $G[a]$ does not change sign on $I \times I$ and $G[a]$ vanishes at some point $(t_0, s_0) \in I \times I$, then either (t_0, s_0) belongs to the diagonal of the square $I \times I$ or (t_0, s_0) is in the boundary of $I \times I$, that is, at least one of the three following properties holds:*

1. *$t_0 = s_0 \in I$.*
2. *$t_0 = 0$ or $t_0 = T$.*
3. *$s_0 = 0$ or $s_0 = T$.*

Proof. Suppose, on the contrary, that $G[a](t_0, s_0) = 0$ for some $(t_0, s_0) \in (0, T) \times (0, T)$ such that $t_0 \neq s_0$. Since $G[a](t_0, s_0) = G[a](s_0, t_0)$, we may assume $t_0 > s_0$.

By definition of the Green's function, we know that

$$x(t) \equiv G[a](t, s_0), \qquad t \in I,$$

solves the equation

$$x''(t) + a(t)\, x(t) = 0, \quad \text{a.e. } t \in (s_0, T], \quad x(t_0) = x'(t_0) = 0.$$

Then, $G[a](t, s_0) = 0$ for all $t \in (s_0, T]$ and, in consequence, from the symmetric property, $G[a](s_0, s) = 0$ for all $s \in (s_0, T]$.

Now, fix $s \in (s_0, T]$, since $G[a]$ is nonnegative on $I \times I$, we have that function

$$y(t) \equiv G[a](t, s), \qquad t \in I,$$

is a solution of

$$y''(t) + a(t)\, y(t) = 0, \quad \text{a.e. } t \in [0, s), \quad y(s_0) = y'(s_0) = 0.$$

Once again, $G[a](t, s) = 0$ for all $s \in (s_0, T]$ and all $t \in [0, s)$.

From symmetry, we deduce $G[a](t, s) = 0$ for all $t \in (s_0, T]$ and $s \in [0, t)$. This contradicts (G3) and so we deduce the result. ☐

Remark 15. If we consider the periodic case with $a(t) = \left(\frac{\pi}{T}\right)^2$, using [8] we obtain the following expression for the Green's function

$$G_P[a](t, s) = \frac{T}{2\pi} \begin{cases} \sin\left(\frac{\pi\,(t-s)}{T}\right), & 0 \leq s \leq t \leq T, \\ \sin\left(\frac{\pi\,(t-s+T)}{T}\right), & 0 \leq t < s \leq T. \end{cases}$$

We note that the Green's function is strictly positive on $I \times I$ except for the diagonal and the points $(0, T)$ and $(T, 0)$.

On the other hand, when $a(t) = k^2 < \left(\frac{\pi}{T}\right)^2$ and the Dirichlet boundary conditions are studied, we have that the Green's function is given by the following expression

$$G_D[a](t, s) = \frac{1}{k \sin(k T)} \begin{cases} \sin(k s) \sin(k(t - T)), & 0 \le s \le t \le T, \\ \sin(k t) \sin(k(s - T)), & 0 \le t < s \le T. \end{cases}$$

We observe that $G_D[a]$ is strictly negative on $(0, T) \times (0, T)$ and vanishes on the boundary of its square of definition.

In consequence, the previous result cannot be improved for general self-adjoint Hill's operators.

Now, we will obtain the relation between the sign of the Green's function and the comparison principles for $L[a]$. Such result can also be found in [36, Theorem 4.1].

Lemma 10. *The following claims are equivalent:*
(1) $G[a](t, s) \ge 0 \ (\le 0)$ on $I \times I$.
(2) If $u \in X$ and $L[a] u \succ 0$ on I then $u > 0 \ (< 0)$ on $(0, T)$.

Proof. First observe that inequality $L[a] u \succ 0$ on I is equivalent to the existence of some $\sigma \in L^1(I)$ such that $\sigma \succ 0$ on I, for which Eq. (3.4) is fulfilled. If the Green's function does not change sign, we deduce the strict constant sign of u on $(0, T)$ as a direct consequence of (3.5) and Proposition 2.

Reciprocally, assume *(2)* and suppose that $G[a]$ changes sign on $I \times I$. Arguing as in [3, Theorem 3.1], one can find $t_0 \in I$ and $u_1, u_2 \in X$ such that $L[a] u_1 \succ 0$, $L[a] u_2 \succ 0$ on I and $u_1(t_0) u_2(t_0) < 0$.

First we will prove that there exist $t_0 \in (0, T)$, $s_1, s_2 \in I$ such that $G[a](t_0, s_1) > 0$ and $G[a](t_0, s_2) < 0$. On the contrary, if $G[a](t, \cdot)$ has constant sign for all $t \in (0, T)$ then, due to the change of sign of $G[a]$, there will exist some $t_1 \in (0, T)$ such that $G[a](t_1, \cdot) \equiv 0$. From symmetry, $G[a](\cdot, t_1) \equiv 0$, which contradicts the fact that

$$\frac{\partial G[a]}{\partial t}(t_1^+, t_1) - \frac{\partial G[a]}{\partial t}(t_1^-, t_1) = 1.$$

Therefore the existence of such t_0 is ensured.

Then there exists a neighbourhood of s_1, $A_1 \subset [0, T]$, in which $G[a](t_0, \cdot)$ is positive. If we choose a function f_1 which is positive on

A_1 and vanishes on $[0, T] \setminus A_1$, we deduce that there exists u_1 satisfying $L[a] u_1 = f_1 \succ 0$ and

$$u_1(t_0) = \int_{A_1} G[a](t_0, s) f_1(s) \, ds > 0.$$

Analogously, there exists a neighbourhood of s_2, $A_2 \subset [0, T]$, in which $G[a](t_0, \cdot)$ is negative. Choosing now f_2 which is positive on A_2 and vanishes on $[0, T] \setminus A_2$, we have that there exists u_2 such that $L[a] u_2 = f_2 \succ 0$ and

$$u_2(t_0) = \int_{A_2} G[a](t_0, s) f_2(s) \, ds < 0.$$

Therefore we reach a contradiction with *(2)*. □

As we have noted in Remark 14, the strict constant sign in *(2)* could also be extended to 0 and (or) T, depending on the boundary conditions involved in X. This way, denoting by G_P, G_N, G_{M_1}, and G_{M_2} the corresponding Green's functions related to operator $L[a]$ on the spaces X_P, X_N, X_{M_1}, and X_{M_2}, respectively, we have the following equivalences.

Lemma 11. *The following claims are equivalent:*
(1) $G_P[a](t, s) \geq 0 \ (\leq 0)$ *on $I \times I$.*
(2) *If $u \in X_P$ and $L[a] u \succ 0$ on I then $u > 0 \ (< 0)$ on I.*

Lemma 12. *The following claims are equivalent:*
(1) $G_N[a](t, s) \geq 0 \ (\leq 0)$ *on $I \times I$.*
(2) *If $u \in X_N$ and $L[a] u \succ 0$ on I then $u > 0 \ (< 0)$ on I.*

Lemma 13. *The following claims are equivalent:*
(1) $G_{M_1}[a](t, s) \geq 0 \ (\leq 0)$ *on $I \times I$.*
(2) *If $u \in X_{M_1}$ and $L[a] u \succ 0$ on I then $u > 0 \ (< 0)$ on $[0, T)$.*

Lemma 14. *The following claims are equivalent:*
(1) $G_{M_2}[a](t, s) \geq 0 \ (\leq 0)$ *on $I \times I$.*
(2) *If $u \in X_{M_2}$ and $L[a] u \succ 0$ on I then $u > 0 \ (< 0)$ on $(0, T]$.*

We finalize this preliminary section by showing two particular cases of some more general results given in [4] (Lemmas 1.8.25 and 1.8.33).

Lemma 15. *Suppose that operator $L[a]$ is nonresonant in the Banach space X, its related Green's function $G[a]$ is nonpositive on $I \times I$, and satisfies condition*

(N_g)   There is a continuous function $\phi(t) > 0$ for all $t \in (0, T)$ and k_1, $k_2 \in$
    $L^1(I)$, such that $k_1(s) < k_2(s) < 0$ for a.e. $s \in I$, satisfying

$$\phi(t) k_1(s) \leq G[a](t, s) \leq \phi(t) k_2(s), \quad \text{for a.e. } (t, s) \in I \times I.$$

Then the Green's function $G[a + \lambda]$ is nonpositive on $I \times I$ if and only if $\lambda \in (-\infty, \lambda_0[a])$ or $\lambda \in [-\bar{\mu}(a), \lambda_0[a])$, with $\lambda_0[a] > 0$ the first eigenvalue of operator $L[a]$ in X and $\bar{\mu}(a) \geq 0$ such that $L[a - \bar{\mu}(a)]$ is nonresonant in X and the related nonpositive Green's function $G[a - \bar{\mu}(a)]$ vanishes at some point of the square $I \times I$.

Lemma 16. Suppose that operator $L[a]$ is nonresonant in the Banach space X, its related Green's function $G[a]$ is nonnegative on $I \times I$, and satisfies condition
(P_g)   There is a continuous function $\phi(t) > 0$ for all $t \in (0, T)$ and k_1, $k_2 \in$
    $L^1(I)$, such that $0 < k_1(s) < k_2(s)$ for a.e. $s \in I$, satisfying

$$\phi(t) k_1(s) \leq G[a](t, s) \leq \phi(t) k_2(s), \quad \text{for a.e. } (t, s) \in I \times I.$$

Then the Green's function $G[a + \lambda]$ is nonnegative on $I \times I$ if and only if $\lambda \in (\lambda_0[a], \infty)$ or $\lambda \in (\lambda_0[a], \bar{\mu}(a)]$, with $\lambda_0[a] < 0$ the first eigenvalue of operator $L[a]$ in X and $\bar{\mu}(a) \geq 0$ such that $L[a + \bar{\mu}(a)]$ is nonresonant in X and the related nonnegative Green's function $G[a + \bar{\mu}(a)]$ vanishes at some point of the square $I \times I$.

It is obvious that if the Green's function is strictly positive (strictly negative) on $I \times I$ then condition (P_g) ((N_g)) is trivially fulfilled.

3.3 PERIODIC CONDITIONS

3.3.1 Properties of the Periodic Green's Function

This section is devoted to the study of some useful properties of the Green's function related to operator $L[a]$ coupled with the periodic boundary conditions. Note that in this case the Green's function is symmetric.

Let now $s \in I$ be given. From the definition of the Green's function, using the T-periodicity of the function a, and denoting by

$$u_s(\cdot) = G_P[a](\cdot, s), \tag{3.6}$$

it is not difficult to verify that $u_s \in W^{2,1}(I_s^k)$, where

$$I_s^k = (s + kT, s + (k + 1) T),$$

and it satisfies the equation

$$u_s''(t) + a(t)\, u_s(t) = 0, \text{ a.e. } t \in I_s^k, \quad k \in \mathbb{Z}. \tag{3.7}$$

Because of this property, it is possible to improve Proposition 2 for the periodic case.

Lemma 17. *Suppose that the Green's function $G_P[a]$ does not change sign on $I \times I$ and $G_P[a]$ vanishes at some point $(t_0, s_0) \in I \times I$, then $t_0 = s_0$, $(t_0, s_0) = (0, T)$ or $(t_0, s_0) = (T, 0)$.*

Proof. Suppose, on the contrary, that $G_P[a](t_0, s_0) = 0$ for some $(t_0, s_0) \neq (0, T)$, $(T, 0)$ such that $t_0 \neq s_0$. Since $G_P[a](t_0, s_0) = G_P[a](s_0, t_0)$, we may assume $t_0 > s_0$. From Eq. (3.7), we know that function

$$u(t) \equiv G_P[a](t, s_0), \quad t \in \mathbb{R},$$

solves the equation

$$u''(t) + a(t)\, u(t) = 0, \quad \text{a.e. } t \in (s_0, s_0 + T), \quad u(t_0) = u'(t_0) = 0.$$

So, $G_P[a](t, s_0) = 0$ for all $t \in (s_0, s_0 + T)$. But this contradicts the fact that

$$\frac{\partial G_P[a]}{\partial t}(t^+, t) - \frac{\partial G_P[a]}{\partial t}(t^-, t) = 1, \quad \text{for all } t \in \mathbb{R}. \tag{3.8}$$

□

Remark 16. If we consider the constant potential $a(t) \equiv \left(\frac{\pi}{T}\right)^2$, we have seen in Remark 15 that the Green's function is strictly positive on $I \times I$ except at the diagonal and at the points $(0, T)$ and $(T, 0)$. In consequence the previous result is optimal.

Moreover, it is easy to see that

Lemma 18. *The four following equalities are equivalent:*
1. $G_P[a](T, 0) = 0$.
2. $G_P[a](0, T) = 0$.
3. $G_P[a](0, 0) = 0$.
4. $G_P[a](T, T) = 0$.

Proof. Suppose that $G_P[a](T, 0) = 0$ then, by symmetry we deduce that $G_P[a](0, T) = 0$. The rest of the proof follows from the periodicity of the Green's function. □

Therefore, we arrive at the following conclusion.

Corollary 5. *If the Green's function $G_P[a]$ does not change sign on $I \times I$, then, for all $t \in I$ the functions $G_P[a](t, \cdot)$ and $G_P[a](\cdot, s)$ vanish on I in, at most, two points. Even more, in such a case we have:*
1. *If $s_0 = 0$ or $s_0 = T$, then $G_P[a](\cdot, s_0)$ can only vanish at $t = 0$ and $t = T$.*
2. *If $t_0 = 0$ or $t_0 = T$, then $G_P[a](t_0, \cdot)$ can only vanish at $s = 0$ and $s = T$.*
3. *If $s_0 \in (0, T)$, then $G_P[a](\cdot, s_0)$ can only vanish at $t = s_0$.*
4. *If $t_0 \in (0, T)$, then $G_P[a](t_0, \cdot)$ can only vanish at $s = t_0$.*

As a consequence of the two previous results, we deduce the following property for nonpositive Green's functions.

Lemma 19. *If $G_P[a](t, s) \leq 0$ on $I \times I$ then $G_P[a](t, s) < 0$ on $I \times I$.*

Proof. From Lemmas 17 and 18 we have two possibilities:
1. *There exists $t_0 \in (0, T)$ such that $G_P[a](t_0, t_0) = 0$.*
   Since $G_P[a]$ is nonpositive, we know, from (G3), that

$$\frac{\partial G_P[a]}{\partial t}(t_0^+, t_0) = \frac{\partial G_P[a]}{\partial t}(t_0^-, t_0) + 1 \geq 1,$$

   which implies that $G_P[a]$ is positive on a right-neighbourhood of t_0, and we attain a contradiction.
2. $G_P[a](0, 0) = 0$.
   From (3.7), we have that

$$u_0''(t) + a(t)\, u_0(t) = 0, \quad \text{a.e. } t \in I_0^k, \quad k \in \mathbb{Z},$$

   and from (3.8) we obtain

$$u_0'(0^+) = u_0'(0^-) + 1 = u_0'(T^-) + 1.$$

On the other hand, Lemma 18 implies that $u_0(0) = u_0(T) = 0$ and the nonpositiveness of u gives us

$$u_0'(0^+) \leq 0 \leq u_0'(T^-) < u_0'(0^+),$$

which is a contradiction.

$\square$

From the definition of u_s given in (3.7), and reasoning as in the previous lemmas, we deduce the following result.

Lemma 20. *Suppose that the Green's function $G_P[a]$ does not change sign on $I \times I$. Then it is nonnegative on $I \times I$ and vanishes at some point $(t_0, t_0) \in I \times I$ if and only if the equation*

$$u''(t) + a(t)\, u(t) = 0, \quad t \in (t_0, t_0 + T), \quad u(t_0) = u(T + t_0) = 0$$

has a nontrivial constant sign solution.

Defining now the function

$$a_s(t) \equiv a(t + s), \quad s, t \in \mathbb{R},$$

we arrive at the following result.

Lemma 21. *For all $t, s \in \mathbb{R}$, we have $G_P[a](t, s) = G_P[a_s](t - s, 0)$.*

Proof. From the periodicity of $G_P[a]$ and condition (3.8), we have that $u_s(\cdot) := G_P[a](\cdot, s)$ is the unique solution of the equation

$$u_s''(t) + a(t)\, u_s(t) = 0, \quad \text{a.e. } t \in (s, s + T),$$
$$u_s(s) = u_s(s + T), \quad u_s'(s^+) = u_s'((s + T)^-) + 1.$$

On the other hand, $y_s(t) := u_s(t + s)$ is the unique solution of the equation

$$y_s''(t) + a(t + s)\, y_s(t) = 0, \quad \text{a.e. } t \in (0, T),$$
$$y_s(0) = y_s(T), \quad y_s'(0^+) = y_s'(T^-) + 1.$$

As a consequence, $u_s(s + t) = G_P[a_s](t, 0)$ or, which is the same, $G_P[a](t, s) \equiv u_s(t) = G_P[a_s](t - s, 0)$. □

Remark 17. We notice that the previous property extends to a nonconstant potential a the expression obtained in [3, Lemma 2.1] for constant ones, which implies that if a is constant then the Green's function $G_P[a]$ is constant along the straight lines of slope equal to one.

Lemma 21 allows us to rewrite Lemma 20 as follows.

Corollary 6. *Suppose that the Green's function $G_P[a]$ does not change sign on $I \times I$. Then it vanishes at some point $(t_0, t_0) \in I \times I$ if and only if the equation*

$$u''(t) + a_{t_0}(t)\, u(t) = 0, \quad t \in I, \quad u(0) = u(T) = 0,$$

has a nontrivial constant sign solution.

Moreover, we deduce the following result.

Lemma 22. *Let* $b(t) \equiv a(T - t)$ *for all* $t \in I$. *Then the following equality holds:*

$$G_P[a](t, s) = G_P[b](T - t, T - s) \quad \text{for all } t, s \in I.$$

Proof. We know that $u(t) := G_P[a_s](t, 0)$ is the unique solution of equation

$$u''(t) + a(t + s) u(t) = 0, \quad \text{a.e. } t \in (0, T),$$
$$u(0) = u(T), \quad u'(0^+) = u'(T^-) + 1.$$

So, $y(t) = u(T - t)$ is the unique solution of equation

$$y''(t) + b(t + s) y(t) = 0, \quad \text{a.e. } t \in (0, T),$$
$$y(0) = y(T), \quad y'(0^+) = y'(T^-) + 1,$$

that is, $y(t) := G_P[a_s](T - t, 0) = G_P[b_s](t, 0)$.

Now, from Lemma 21 and the properties of the Green's function, we deduce that

$$G_P[a](t, s) = G_P[a_s](t - s, 0) = G_P[b_s](T - t + s, 0) = G_P[b](T - t, -s)$$
$$= G_P[b](T - t, T - s). \qquad \square$$

As a consequence of the two previous lemmas we can deduce the following corollary.

Corollary 7. *Let* a_s *and* b_r *be defined as in the two previous lemmas for some* $s, r \in I$, *then the functions* $G_P[a_s]$ *and* $G_P[b_r]$ *take exactly the same values (at different points) on* $I \times I$.

To finish this section, we obtain the following comparison results for the Green's functions related to different potentials.

Lemma 23. *Let* $a_1, a_2 \in L^\alpha(I)$ *be such that the corresponding related Green's functions of the periodic problem,* $G_P[a_1]$ *and* $G_P[a_2]$, *do not change sign on* $I \times I$ *and* $G_P[a_1](t, s) \, G_P[a_2](t, s) \geq 0$ *for all* $(t, s) \in I \times I$. *If* $a_1 \succ a_2$ *on* I *then* $G_P[a_1](t, s) < G_P[a_2](t, s)$ *for all* $(t, s) \in I \times I$.

Proof. Let $s \in I$ be given. Denote by $u_s(\cdot) = G_P[a_1](\cdot, s)$ and $y_s(\cdot) = G_P[a_2](\cdot, s)$. From the definition of the Green's functions, we know that

$$u_s''(t) + a_1(t) u_s(t) = y_s''(t) + a_2(t) y_s(t) = 0, \quad \text{a.e. } t \in (s, s + T),$$

and

$$u_s(s) = u_s(s+T), \quad u_s'(s^+) = u_s((s+T)^-) + 1,$$
$$y_s(s) = y_s(s+T), \quad y_s'(s^+) = y_s'((s+T)^-) + 1.$$

Now, for all $t \in I$, we define the functions $\bar{u}_s(t) := u_s(t+s)$ and $\bar{y}_s(t) := y_s(t+s)$. Therefore

$$\bar{u}_s - \bar{y}_s \in X_P.$$

Suppose now that the Green's functions are nonnegative (the other case is analogous). In consequence, from Lemma 11 it follows that $\bar{u}_s > 0$ and $\bar{y}_s > 0$ for all $t \in I$.

Let $\epsilon(t) = a_1(t) - a_2(t) \succ 0$ on I. We have that

$$(\bar{y}_s - \bar{u}_s)''(t) + a_1(t+s)(\bar{y}_s - \bar{u}_s)(t) = \epsilon(t+s)\bar{y}_s(t) \succ 0, \quad \text{a.e. } t \in (0, T),$$

and we deduce, from Lemma 11, that $\bar{y}_s > \bar{u}_s$ on I, or, which is the same, $G_P[a_1](t, s) < G_P[a_2](t, s)$ for all $(t, s) \in I \times I$. $\qquad\square$

3.3.2 Optimal Conditions for the Periodic MP and AMP

We will show now a characterization of the comparison principles due to M. Zhang by using the corresponding eigenvalues of the related homogeneous equation. Such characterization allows us to describe the constant sign of the Green's functions from the oscillation properties of operator $L[a]$.

Let $\lambda_0^P[a]$ be the smallest eigenvalue of the periodic equation

$$u''(t) + (a(t) + \lambda)u(t) = 0, \quad \text{a.e. } t \in I, \quad u(0) = u(T), \; u'(0) = u'(T),$$

and $\lambda_0^A[a]$ the smallest eigenvalue of the antiperiodic equation

$$u''(t) + (a(t) + \lambda)u(t) = 0, \quad \text{a.e. } t \in I, \quad u(0) = -u(T), \; u'(0) = -u'(T).$$

We have proved in Chapter 2 (Theorem 11) that $\lambda_0^P[a] < \lambda_0^A[a]$. In [36], M. Zhang obtained the following result (in the paper for $T = 1$).

Lemma 24. *[36, Theorem 1.1] Suppose that $a \in L^1(I)$, then:*
1. *$L[a]$ admits MP if and only if $\lambda_0^P[a] > 0$.*
2. *$L[a]$ admits AMP if and only if $\lambda_0^P[a] < 0 \leq \lambda_0^A[a]$.*

By introducing the parametrized potentials $a + \lambda$ the previous result can be rewritten as follows.

Lemma 25. *[36, Theorem 1.2] Suppose that $a \in L^1(I)$, then:*
1. $L[a + \lambda]$ *admits MP if and only if* $\lambda < \lambda_0^P[a]$.
2. $L[a + \lambda]$ *admits AMP if and only if* $\lambda_0^P[a] < \lambda \leq \lambda_0^A[a]$.

The following explicit bounds for the first periodic and antiperiodic eigenvalues are well-known.

Lemma 26. *Suppose that $a \in L^1(I)$, then:*
(i) $\lambda_0^P[a] \leq -\frac{1}{T}\int_0^T a(s)ds$ *and the equality holds if and only if a is constant.*
(ii) $\lambda_0^A[a] = \min\{\lambda_0^D[a_s], s \in \mathbb{R}\}$, *where $\lambda_0^D[a_s]$ is the first eigenvalue of the Dirichlet problem*

$$u''(t) + (a_s(t) + \lambda)u(t) = 0, \quad a.e. \ t \in I, \quad u(0) = u(T) = 0. \quad (3.9)$$

(iii) *If $\|a_+\|_\alpha \leq K(2\alpha^*, T)$, then $\lambda_0^A[a] \geq \left(\frac{\pi}{T}\right)^2 \left(1 - \frac{\|a_+\|_\alpha}{K(2\alpha^*, T)}\right) \geq 0.$*

Proof. (i) Let u be a nontrivial T-periodic solution associated to $\lambda_0^P[a]$, that is,

$$u''(t) + \left(a(t) + \lambda_0^P[a]\right)u(t) = 0, \quad t \in I, \quad u(0) = u(T), \quad u'(0) = u'(T).$$

By the results we have seen in Chapter 2 (Theorem 12 and Remark 13) we know that u does not vanish. Then, we can define $h(t) = \dfrac{u'(t)}{u(t)}$, which satisfies $h(0) = h(T)$ and

$$h'(t) = -\left(\lambda_0^P[a] + a(t)\right) - h^2(t).$$

Integrating in $[0, T]$ we get

$$\lambda_0^P[a] + \frac{1}{T}\int_0^T a(s)ds = -\frac{1}{T}\int_0^T h^2(s)ds.$$

Therefore, $\lambda_0^P[a] \leq -\frac{1}{T}\int_0^T a(s)ds$ and the equality holds if and only if $h \equiv 0$, which means that u is constant. This is equivalent to a being constant too.

(ii) See [37].

(iii) From the variational characterization of eigenvalues it is well-known that

$$\lambda_0^D[a] = \min_{\|u\|_2 = 1} q(u), \quad u \in H_0^1(I),$$

where q is the quadratic form

$$q(u) = \int_0^T (u')^2(s)\, ds - \int_0^T a(s)\, u^2(s)\, ds.$$

Then, if $u \in H_0^1(I)$ with $\|u\|_2 = 1$, from Hölder's inequality and the definition of the Sobolev constant K, introduced in (3.3), we have

$$q(u) \geq \|u'\|_2^2 - \|a_+\|_\alpha \|u^2\|_{\alpha^*} = \|u'\|_2^2 - \|a_+\|_\alpha \|u\|_{2\alpha^*}^2$$

$$\geq \left(1 - \frac{\|a_+\|_\alpha}{K(2\alpha^*, T)}\right) \|u'\|_2^2$$

$$\geq \left(1 - \frac{\|a_+\|_\alpha}{K(2\alpha^*, T)}\right) K(2, T) = \left(1 - \frac{\|a_+\|_\alpha}{K(2\alpha^*, T)}\right)\left(\frac{\pi}{T}\right)^2 \geq 0.$$

Thus, $\lambda_0^D[a] \geq \left(1 - \dfrac{\|a_+\|_\alpha}{K(2\alpha^*, T)}\right)\left(\dfrac{\pi}{T}\right)^2 \geq 0$. Now, taking into account (ii) and the fact that $\|(a_s)_+\|_\alpha = \|a_+\|_\alpha$, one obtains

$$\lambda_0^A[a] \geq \left(\frac{\pi}{T}\right)^2 \left(1 - \frac{\|a_+\|_\alpha}{K(2\alpha^*, T)}\right) \geq 0. \qquad \square$$

As a direct consequence of item (*ii*) in the previous lemma and Lemma 2 and Remark 13 in Chapter 2,

Lemma 27. *The function*

$$\lambda_0^A: L^\alpha(I) \to \mathbb{R}$$

$$a \mapsto \lambda_0^A[a]$$

that assigns to every potential a the first eigenvalue of the corresponding antiperiodic problem, is decreasing.

An analogous result can be proved for the periodic problem.

Lemma 28. *The function*

$$\lambda_0^P: L^\alpha(I) \to \mathbb{R}$$

$$a \mapsto \lambda_0^P[a]$$

that assigns to every potential a the first eigenvalue of the corresponding periodic problem, is decreasing.

Proof. Let $a \in L^\alpha(I)$. As we have seen in Chapter 2 (Theorem 12 and Remark 13), the eigenfunction u_0^P related to the first eigenvalue of the periodic problem has constant sign (which can be chosen positive), that is, $u_0^P > 0$ is a solution of the following problem

$$u''(t) + \left(a(t) + \lambda_0^P[a]\right) u(t) = 0, \quad t \in [0, T],$$
$$u(0) = u(T), \quad u'(0) = u'(T).$$

Let now $\varepsilon \in L^\alpha(I)$, $\varepsilon \succ 0$. Then

$$\left(u_0^P\right)''(t) + \left(a(t) + \varepsilon(t) + \lambda_0^P[a]\right) u_0^P(t) = \varepsilon(t) u_0^P(t) \succ 0, \quad t \in [0, T],$$
$$u_0^P(0) = u_0^P(T), \quad \left(u_0^P\right)'(0) = \left(u_0^P\right)'(T).$$

Therefore it is clear that operator $L[\lambda_0^P[a] + a + \varepsilon]$ does not satisfy the MP and, from Lemma 25, this is equivalent to

$$\lambda_0^P[a + \varepsilon] \le \lambda_0^P[a].$$

Thus we conclude that λ_0^P is a decreasing function.   □

Now, as a consequence of the previous lemmas and Lemma 23, we arrive at the following result.

Corollary 8. *Let $a_1, a_2 \in L^\alpha(I)$ be such that $a_1 \succ a_2$ on I and assume that the related Green's functions, $G_P[a_1]$ and $G_P[a_2]$, have the same constant sign on $I \times I$. Then operator $L[a]$ is nonresonant in X_P and $G_P[a]$ has the same constant sign for all $a \in L^\alpha(I)$ such that $a(t) \in [a_2(t), a_1(t)]$ for a.e. $t \in I$.*

Proof. Let $a \in L^\alpha(I)$ be such that $a(t) \in [a_2(t), a_1(t)]$ for a.e. $t \in I$. Then, from Lemmas 27 and 28, we have that

$$\lambda_0^P[a_1] \le \lambda_0^P[a] \le \lambda_0^P[a_2] \quad \text{and} \quad \lambda_0^A[a_1] \le \lambda_0^A[a] \le \lambda_0^A[a_2].$$

Therefore if $G_P[a_1]$ and $G_P[a_2]$ are negative, we deduce from Lemma 24 that

$$0 < \lambda_0^P[a_1] \le \lambda_0^P[a],$$

and consequently $G_P[a]$ is negative.

The same way, if $G_P[a_1]$ and $G_P[a_2]$ are nonnegative, Lemma 24 implies that

$$\lambda_0^P[a] \le \lambda_0^P[a_2] < 0 \le \lambda_0^A[a_1] \le \lambda_0^A[a],$$

and so $G_P[a]$ is nonnegative.   □

As a consequence of the results showed in this section we arrive at the following characterization of the sign of the Green's function.

Theorem 15. *[6, Theorem 3.1] Let R be the infimum of the distance of two consecutive zeros of a solution of the equation $L[a] u = 0$. Then the following assertions hold:*

1. *$G_P[a]$ changes sign in $I \times I$ if and only if $R < T$.*
2. *$G_P[a]$ is nonnegative and vanishes at some points on $I \times I$ if and only if $R = T$.*
3. *$G_P[a]$ has strict constant sign in $I \times I$ if and only if $R > T$.*

Proof. If $R < T$, we have that there is $s \in I$ for which at least one solution of the equation $u''(t) + a_s(t) u(t) = 0$ has two zeros in I. From Chapter 2, we have that $\lambda_0^D[a_s]$, the first eigenvalue of the Dirichlet problem

$$L[a_s] u = 0 \text{ on } I, \quad u(0) = u(T) = 0,$$

is strictly negative. In consequence, Lemma 26, *(ii)*, implies that $\lambda_0^A[a] < 0$. Now, Lemma 24 ensures that $G_P[a]$ changes sign on $I \times I$.

When $R = T$, we conclude, as above, that $\lambda_0^A[a] = 0$. So Lemma 24 says us that $G_P[a]$ is nonnegative on $I \times I$. So, Corollary 6 and Lemma 26 *(ii)* show that $G_P[a]$ vanishes at some points on $I \times I$.

When $R > T$ we will follow the argument made in [30, Theorem 2.1]. Since $G_P[a]$ is a continuous function on $I \times I$, it will be enough to prove that it does not vanish at any point. By contradiction, suppose that there exists $(t_0, s_0) \in I \times I$ such that $G_P[a](t_0, s_0) = 0$. First, assume that $t_0 \neq 0$, T and consider a fixed $s_0 \in (0, T)$.

Suppose that $t_0 \geq s_0$. Then $u_{s_0}(\cdot) := G_P[a](\cdot, s_0)$ (defined in (3.6)) is a solution of $L[a] u = 0$ on the interval $I_{s_0}^0 = [s_0, s_0 + T]$ and satisfies that $u_{s_0}(t_0) = 0$. Since, by hypothesis, two consecutive zeros must be separated by a distance greater than T, u_{s_0} can not cancel at any other point in the interval $[s_0, s_0 + T]$. In particular, $u_{s_0}(s_0) = u_{s_0}(s_0 + T) \neq 0$ and as a consequence, t_0 must be either a maximum or a minimum of u_{s_0} in $[s_0, s_0 + T]$, that is, $u'_{s_0}(t_0) = 0$.

By the uniqueness of solution of the initial value problem, we have that $u_{s_0}(t) \equiv 0$ for all t, which is a contradiction with the properties of the Green's function as, for instance, the fact that

$$\frac{\partial G_P[a]}{\partial t}(t^+, t) = \frac{\partial G_P[a]}{\partial t}(t^-, t) + 1, \quad \forall t \in I.$$

Analogously, if $t_0 < s_0$ we could consider the function u_{s_0} as a solution of $L[a]u = 0$ on the interval $I_{s_0}^{-1} = [s_0 - T, s_0]$ and follow the same reasoning.

Finally, if either $s_0 = 0$ or $s_0 = T$, then $G_P[a](t, s_0)$ is a solution of $L[a]u = 0$ in I and satisfies that $G_P[a](0, s_0) = G_P[a](T, s_0)$ and following the same argument as before we would reach a contradiction.

Moreover, the cases $t_0 = 0$ and $t_0 = T$ are covered as a consequence of the symmetry of $G_P[a]$.

Therefore we conclude that $G_P[a]$ can not vanish on $I \times I$ and so it has constant sign. □

3.3.3 Explicit Criteria for the Periodic AMP and MP

As a consequence of Lemmas 24 and 26 we have the following explicit criteria for the AMP in the periodic case.

Lemma 29. *Assume that* $\int_0^T a(t)\,dt \geq 0$, $a \not\equiv 0$, *and moreover*

$$\|a_+\|_\alpha \leq K(2\alpha^*, T). \tag{3.10}$$

Then $L[a]$ satisfies the AMP on X_P.

Proof.

Claim. *The distance between two consecutive zeros of a nontrivial solution of* $u''(t) + a(t)\,u(t) = 0$ *is strictly greater that T.*

Assume on the contrary that u is a nontrivial solution of the Dirichlet problem

$$u''(t) + \check{a}(t)\,u(t) = 0,\ t \in [t_1, t_2],\quad u(t_1) = 0 = u(t_2), \tag{3.11}$$

where $0 < t_2 - t_1 \leq T$ and $\check{a}$ is the restriction of function a to the interval $[t_1, t_2]$.

It is clear, from expression (3.3), that for any fixed α, the expression $K(\alpha, T)$ is strictly decreasing in $T > 0$. Then, since $0 < t_2 - t_1 \leq T$, we deduce the following properties:

$$\|\check{a}_+\|_p \leq \|a_+\|_p < K(2p^*, T) \leq K(2p^*, t_2 - t_1).$$

From Lemma 26, (iii), it follows that

$$\lambda_0^D[\check{a}] > 0,$$

which contradicts that (3.11) has a nontrivial solution.

Now, Theorem 15 and the *Claim* imply that $G_P[a]$ has strict constant sign. To determinate its sign, consider the periodic problem

$$u''(t) + a(t)\,u(t) = 1, \quad u(0) = u(T), \quad u'(0) = u'(T). \tag{3.12}$$

It is clear that its unique solution is given by the expression

$$u(t) = \int_0^T G_P[a](t, s)\, ds. \tag{3.13}$$

Obviously u does not vanish and has the same sign as $G_P[a]$. Then, dividing the equation by u and integrating over I we obtain

$$0 < \int_0^T \left(\frac{u'(t)}{u(t)} \right)^2 dt + \int_0^T a(t)\, dt = \int_0^T \frac{dt}{u(t)}.$$

Hence $u(t) > 0$ on I which implies that $G_P[a] > 0$ on $I \times I$ or, which is the same, $L[a]$ satisfies the AMP on X_P. □

Lemma 29 includes as particular cases [30, Corollary 2.3] and [5, Theorem 3.2] and it was extended in [9, Theorem 3.4 and Remark 3.7] to the p-Laplacian equation

$$\left(|u'|^{p-2}\,u'\right)' + a(t)\left(|u|^{p-2}\,u\right) = h(t), \quad u(0) = u(T), \quad u'(0) = u'(T).$$

We point out that M. Zhang constructs in [36] some examples of potentials a for which $L[a]$ admits the AMP but inequality (3.10) does not hold.

When we refer to the study of MP, the following general result for nonpositive (and not identically zero) potentials was obtained by P. Torres.

Lemma 30. *[30, Corollary 2.2] If $a \prec 0$ then $L[a]$ satisfies the MP on X_P.*

Proof. If $a \prec 0$, from Theorem 7 we know that any nontrivial solution of

$$u''(t) + a(t)\,u(t) = 0$$

has at most one zero. Then, from Theorem 15, the related Green's function $G_P[a]$ has constant sign.

On the other hand, if we integrate Eq. (3.12) on I, we obtain

$$\int_0^T a(t)\,u(t)\, dt = T > 0.$$

Since by hypothesis $a \prec 0$, necessarily $u(t) < 0$ for some t and therefore $G_P[a]$ must be negative at some points. As we have proved that $G_P[a]$ has constant sign, we conclude that it is negative on $I \times I$ or, which is the same, operator $L[a]$ satisfies the MP on X_P.   □

In [5] the authors obtained the following result.

Lemma 31. *[5, Theorem 4.1] Assume that $a \in L^{\alpha}(I)$ is of the form*

$$a(t) = b'(t) - b^2(t), \quad b(0) = b(T), \quad \int_0^T b(s)ds \neq 0, \qquad (3.14)$$

where b is an absolutely continuous function. Then $L[a]$ satisfies the MP on X_P.

Proof. The key idea is to decompose the second order operator

$$L[a] u(t) = u''(t) + a(t) u(t),$$

as two first order operators $L[a] = L_1 \circ L_2$, where

$$L_1 u(t) = u'(t) - b(t) u(t) \quad \text{and} \quad L_2 u(t) = u'(t) + b(t) u(t).$$

The following claim is easily proved by direct integration.

Claim. *The problem $u'(t) + b(t) u(t) = h(t)$, $u(0) = u(T)$, has a unique solution for all $h \in L^1(I)$ if and only if*

$$\int_0^T b(s) \, ds \neq 0.$$

Moreover if $h \succ (\prec) 0$ on I then

$$u(t) \int_0^T b(s) \, ds > (<) 0 \quad \text{for all } t \in I.$$

Now, suppose that $\int_0^T b(s)ds > 0$ (being the other case analogous).

If $L[a] u(t) \succ 0$ on I, $u(0) = u(T)$, $u'(0) = u'(T)$ then $L_1(L_2 u)(t) \succ 0$ on I, with $L_2 u(0) = L_2 u(T)$ and from the *Claim* it follows that $L_2 u < 0$ on I. Now the *Claim* implies again that $u < 0$ on I. This fact is equivalent to the negativeness of the Green's function, which concludes the proof.   □

Remark 18. The main difficulty in order to apply Lemma 31 is to determine when the potential a is of the form (3.14). It is easy to see that if $a \not\equiv 0$ satisfies (3.14) then $\int_0^T a(s)\, ds < 0$, but the converse is false as the following example shows: let $a \in L^\infty(0, 1)$, $a(t) \not\equiv 0$, put $a(t) = \hat{a}(t) + \bar{a}$, where $\hat{a}(t)$ has mean value zero and

$$\bar{a} = \frac{1}{T} \int_0^T a(s)\, ds$$

is its mean value, and let us consider the problem:

$$b'(t) - b^2(t) = \hat{a}(t) + \bar{a}, \quad b(0) = b(1). \tag{3.15}$$

By making the change $u(t) = -b(t)$, problem (3.15) is equivalent to

$$u'(t) + u^2(t) + \hat{a}(t) = -\bar{a}, \quad u(0) = u(1), \tag{3.16}$$

and [24, Corollary 3.1] implies that there exists $s_0 \in \mathbb{R}$ such that
  (i)   for $-\bar{a} < s_0$ problem (3.16) has no solution,
  (ii) for $-\bar{a} = s_0$ problem (3.16) has at least one solution,
  (iii) for $-\bar{a} > s_0$ problem (3.16) has at least two solutions.
Since problem (3.16) has no solution for $\bar{a} = 0$, as it is easy to see by integrating the equation over $[0, 1]$, it follows that $s_0 > 0$. So (i) implies that for mean values satisfying $-s_0 < \bar{a} < 0$ problem (3.16) has no solution, and thus there exist potentials with negative average such that (3.14) has no solution.

Recently, R. Hakl and P. Torres, gave the following criterium to ensure the MP.

Lemma 32. [17, Corollary 2.5] If $a \in L^\alpha(I)$, $a \not\equiv 0$, and moreover

$$\int_0^T a_+(s)\, ds < \frac{4}{T}, \quad \frac{\int_0^T a_+(s)\, ds}{1 - \frac{T}{4} \int_0^T a_+(s)\, ds} \leq \int_0^T a_-(s)\, ds,$$

then $L[a]$ satisfies the MP on X_P.

Under the assumptions of each one of the previous three results it follows that $\int_0^T a(s)\, ds < 0$ and in fact this is a necessary condition.

Proposition 3 (Necessary condition for MP). If $L[a]$ satisfies the MP on X_P then

$$\int_0^T a(s)\, ds < 0.$$

Proof. The result follows from Lemma 24, (1) and Lemma 26, (i).   □

Corollary 9. *Suppose that $\int_0^T a(s)\,ds \geq 0$. Then the two following assertions hold:*

1. *If operator $L[a]$ is nonresonant on X_P then it does not satisfy the MP on X_P.*
2. $\lambda_0^P[a] \leq 0$.

Proof. The first assertion is just Proposition 3. The second one is a direct consequence of Lemma 24, (1), coupled with the first part.   □

3.3.4 More on Explicit Criteria

This subsection is based on [6] and it is devoted to the study of the indefinite potentials for which the MP or the AMP on X_P holds and the conditions of the explicit criteria in the previous section are not satisfied. In particular, we will pay special attention to the situations

$$a \not\equiv 0 \text{ and } \int_0^T a(t)\,dt < 0$$

or/and

$$\|a_+\|_\alpha > K(2\alpha^*, T).$$

To this end, we will use the characterization of eigenvalues given in Oscillation Theorem (Theorem 11):

1. λ is a periodic eigenvalue of problem (3.2) if and only if $\Delta(\lambda) = 2$,
2. λ is an antiperiodic eigenvalue of problem (3.2) if and only if $\Delta(\lambda) = -2$,

with $\Delta(\lambda)$ introduced in Definition 8. In particular, problem (3.2) is nonresonant if and only if $\Delta(0) \neq 2$.

Example 2. Suppose that $a(t) \equiv a \in \mathbb{R}$. In this situation, it is not difficult to verify that

$$u_1(t) = \begin{cases} \cosh\sqrt{-a}\,t, & \text{if } a < 0, \\ 1, & \text{if } a = 0, \\ \cos\sqrt{a}\,t, & \text{if } a > 0, \end{cases} \quad \text{and} \quad u_2(t) = \begin{cases} \dfrac{\sinh\sqrt{-a}\,t}{\sqrt{-a}}, & \text{if } a < 0, \\ t, & \text{if } a = 0, \\ \dfrac{\sin\sqrt{a}\,t}{\sqrt{a}}, & \text{if } a > 0. \end{cases}$$

In consequence

$$\Delta(0) = \begin{cases} 2\cosh\sqrt{-a}\,T, & \text{if } a < 0, \\ 0, & \text{if } a = 0, \\ 2\cos\sqrt{a}\,T, & \text{if } a > 0. \end{cases}$$

So, we deduce the very well-known result that this problem is nonresonant if and only if $a \neq \left(\frac{2n\pi}{T}\right)^2$, for all $n = 0, 1, \ldots$.

Proposition 4. *The following assertions hold:*
1. *If $L[a]$ admits MP on X_P then $\Delta(0) > 2$.*
2. *If $L[a]$ admits AMP on X_P then $\Delta(0) < 2$.*

Proof. 1. From Lemma 24, $L[a]$ admits MP on X_P if and only if $\lambda_P^0[a] > 0$. Moreover, as we have seen in Chapters 1 and 2 (Lemma 1 and Theorem 11), $\Delta(\lambda) > 2$ for $\lambda < \lambda_P^0[a]$, from where the result immediately follows.

2. From Lemma 24, we know that $L[a]$ admits AMP on X_P if and only if $\lambda_P^0[a] < 0 \leq \lambda_A^0[a]$. Again, from Theorem 11 in Chapter 2, it holds that $-2 \leq \Delta(\lambda) < 2$ for $\lambda_P^0[a] < \lambda \leq \lambda_A^0[a]$. Therefore, $\Delta(0) < 2$. □

As a consequence of the proof of the previous result, we deduce the following equivalent characterization of the MP and the AMP properties for operator $L[a]$.

Theorem 16. *The following properties hold:*
1. *$L[a]$ satisfies MP on X_P if and only if $\Delta(\lambda) > 2$ for all $\lambda \leq 0$.*
2. *$L[a]$ satisfies AMP on X_P if and only if $\Delta(\lambda) > -2$ for all $\lambda < 0$ and $\Delta(0) < 2$.*

Proof. 1. From Lemma 24, $L[a]$ admits MP on X_P if and only if $\lambda_0^P[a] > 0$. If $\Delta(\lambda) > 2$ for $\lambda \leq 0$ and taking into account that $\Delta(\lambda_0^P[a]) = 2$, it is clear that $\lambda_0^P[a] > 0$.

Conversely, if $\lambda_0^P[a] > 0$, we have seen in the proof of previous lemma that $\Delta(\lambda) > 2$ for $\lambda < \lambda_P^0[a]$, so we deduce that $\Delta(\lambda) > 2$ for all $\lambda \leq 0$.

2. Now, from Lemma 24, $L[a]$ admits AMP on X_P if and only if $\lambda_0^P[a] < 0 \leq \lambda_0^A[a]$.

Moreover, taking into account the fact that $\lambda_0^P[a]$ and $\lambda_0^A[a]$ are the smallest values satisfying $\Delta(\lambda) = 2$ and $\Delta(\lambda) = -2$, respectively, we deduce the following:

- $\Delta(\lambda) > -2$ for all $\lambda < 0$ if and only if $\lambda_0^A[a] \geq 0$.
- If $\Delta(0) < 2$, then $\lambda_0^P[a] < 0$.
- If $\lambda_0^P[a] < 0 \leq \lambda_0^A[a]$, then $-2 = \Delta(\lambda_0^A[a]) \leq \Delta(0) < \Delta(\lambda_0^P[a]) = 2$.

The result follows immediately from the previous assertions. □

We note that obtaining the explicit expression of $\Delta(\lambda)$ is in general not possible for nonconstant potentials a. However there are a lot of very good computer programs, for instance *Maple*, *Mathematica* or *Maxima*, that allow us to get an approximate numerical expression of this formula and its corresponding roots.

Finally, we present a more suitable criteria to ensure the MP and the AMP on X_P character of operator $L[a]$. We note that these conditions depend on the integral of the potential a.

Theorem 17. *Assume* $\|\hat{a}_+\|_\alpha \leq K(2\alpha^*, T)$*, with* $a(t) = \hat{a}(t) + \lambda$*, where* $\hat{a}$ *has mean value zero and* $\lambda = \dfrac{1}{T} \displaystyle\int_0^T a(s)\, ds$ *is the mean value of* a*. Then*

1. $L[a]$ *satisfies MP on* X_P *if and only if* $\displaystyle\int_0^T a(s)\, ds < 0$ *and* $\Delta(0) > 2$.

2. *If* $\displaystyle\int_0^T a(s)\, ds < 0$ *and* $\Delta(0) < 2$ *then* $L[a]$ *satisfies AMP on* X_P.

3. *If* $L[a]$ *is nonresonant in* X *and* $0 \leq \displaystyle\int_0^T a(s)\, ds \leq \dfrac{\pi^2}{T}\left(1 - \dfrac{\|\hat{a}_+\|_\alpha}{K(2\alpha^*, T)}\right)$ *then* $L[a]$ *satisfies AMP on* X_P.

Proof. Since $\|\hat{a}_+\|_\alpha \leq K(2\alpha^*, T)$ we have by Lemma 26, (*iii*), that $\lambda_0^A[\hat{a}] \geq 0$. Moreover, from Lemma 29 and Lemma 25, (2), we know that $\lambda_0^P[\hat{a}] < 0$.

To prove the first assertion, suppose that $L[a]$ satisfies MP on X_P. Then by Propositions 3 and 4 we obtain that $\int_0^T a(s)\, ds < 0$ and $\Delta(0) > 2$.

Reciprocally, assume $\int_0^T a(s)\, ds < 0$ and $\Delta(0) > 2$. Since $\Delta(0) \neq 2$ we have that $L[a] = L[\hat{a} + \lambda]$ is nonresonant and from $\lambda < 0 \leq \lambda_0^A[\hat{a}]$, it follows, from Lemma 25, that either $L[a]$ admits MP or $L[a]$ admits AMP on X_P, depending if either $\lambda < \lambda_0^P[\hat{a}]$ or $\lambda_0^P[\hat{a}] < \lambda < 0$. Finally, $\Delta(0) > 2$ and Proposition 4 lead us to conclude that $L[a]$ admits MP on X_P.

The second part is deduced by repeating the same argument. From the fact that $\lambda < 0 \leq \lambda_0^A[\hat{a}]$, we have that either $L[a]$ admits either MP or AMP, so $\Delta(0) < 2$ and Proposition 4 imply now that $L[a]$ admits AMP on X_P.

The last assertion is deduced from Lemma 26, (*iii*), and the following inequalities:

$$\lambda_0^P[\hat{a}] < 0 \leq \lambda \leq \left(\frac{\pi}{T}\right)^2 \left(1 - \frac{\|\hat{a}_+\|_\alpha}{K(2\alpha^*, T)}\right) \leq \lambda_0^A[\hat{a}]. \qquad \square$$

Remark 19. 1. The sufficient part in assertion (1) of Theorem 17, that is,

If $L[a]$ satisfies MP on X_P then $\displaystyle\int_0^T a(s)\, ds < 0$ and $\Delta(0) > 2$,

is valid for all a satisfying (3.1). In fact it is a direct consequence of Propositions 3 and 4 in which there are no assumptions on $\|\hat{a}_+\|_\alpha$.

2. In Theorem 17 (1) and (2) it remains open to know what happens when the inequality $\|\hat{a}_+\|_\alpha \leq K(2\alpha^*, T)$ is not fulfilled.

3. The assertion (3) in Theorem 17 is optimal for constant potentials $a(t) \equiv k$. Moreover, we notice that if $\int_0^T a(s)\,ds \geq 0$ then $\|\hat{a}_+\|_\alpha \leq \|a_+\|_\alpha$. As a consequence, Theorem 17 (3) can cover wider situations than Lemma 29.

3.3.5 Examples

In this section we present some illustrative examples collected from [6].

Example 3. Consider the problem

$$u''(t) + a_r(t)\, u(t) = h(t), \quad \text{for all } t \in [0, 2], \quad u(0) = u(2), \quad u'(0) = u'(2),$$

with

$$a_r(t) = \begin{cases} -1, & \text{if } 0 \leq t < 1, \\ r, & \text{if } 1 \leq t \leq 2, \end{cases}$$

and $r \in \mathbb{R}$.

Of course, for all $r \leq 0$, we have that $a_r \prec 0$ and, as a consequence, operator $L[a_r]$ satisfies the MP on X_P.

Now, let $r > 0$. In this case $\int_0^2 a_r(t)\,dt = r - 1$ and we can check with this family of potentials the available explicit criteria for MP. Lemma 30 does not apply because a_r is sign-changing. On the other hand, we were not able to verify if a_r is of the form (3.14), so Lemma 31 is not useful for us in this case. Finally Lemma 32 reads as:

- If $0 < r < \frac{2}{3}$ then $L[a_r]$ satisfies the MP on X_P.

Now we compare with Theorem 17. We have that

$$\hat{a}_r(t) = \begin{cases} -\frac{r+1}{2}, & \text{if } 0 \leq t < 1, \\ \frac{r+1}{2}, & \text{if } 1 \leq t \leq 2. \end{cases}$$

Therefore Theorem 17 is applicable whenever

$$\|(\hat{a}_r)_+\|_\alpha = \frac{r+1}{2} \leq \max_{\alpha \geq 1}\left\{K(2\alpha^*, 2)\right\} \approx 2.8125,$$

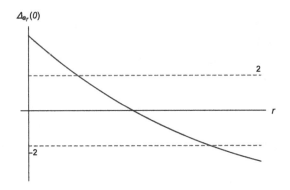

Figure 3.1 Graphic of $\Delta_{a_r}(0)$.

that is

$$0 < r \leq \bar{r} \approx 4.625.$$

After standard computations one can verify that

$$\Delta_{a_r}(0) = \frac{\left(e^2 - 1\right)(1 - r)\sin(\sqrt{r}) + 2\left(1 + e^2\right)\sqrt{r}\cos(\sqrt{r})}{2\,e\,\sqrt{r}},$$

which is graphically represented in Fig. 3.1. Here, Δ_{a_r} denotes the discriminant related to Hill's equation with potential a_r.

So, for $r_0 \approx 0.85724$ we have that $\Delta_{a_{r_0}}(0) = 2$ and $\Delta_{a_r}(0) > 2$ for $0 < r < r_0$. As a consequence, we deduce from Theorem 17 the following properties:

- If $0 < r < r_0$ then $L[a_r]$ satisfies the MP on X_P.
- If $r_0 < r < 1$ then $L[a_r]$ satisfies the AMP on X_P.

Notice that the provided information for the MP case is optimal. Indeed, from Proposition 3 we have that a necessary condition for $L[a_r]$ to satisfy the MP is that $r < 1$. Since we have the AMP when $r_0 < r < 1$ then there are no other possibilities for the MP, that is

- $L[a_r]$ satisfies the MP on X_P if and only if $0 < r < r_0$.

On the other hand, we obtain the following estimation for $r \geq 1$ in order to obtain the AMP,

$$1 \leq r \leq \max_{\alpha \geq 1} \left\{ \frac{\left(1 + \frac{\pi^2}{2} - \frac{\pi^2}{K(2\,\alpha^*,2)}\right)}{\left(1 + \frac{\pi^2}{K(2\,\alpha^*,2)}\right)} \right\} \approx 2.69403.$$

However, this estimation is not the best possible and in fact the bound given in Lemma 29 is better than the previous one:

$$1 \le r \le \max_{\alpha \ge 1} \left\{ K(2\alpha^*, 2) \right\} \approx 2.8125.$$

Moreover, since $\Delta_{a_r}(0) = -2$ for $r = r_1 \approx 3.13363$ and $\Delta_{a_r}(0) \in (-2, 2)$ for all $r \in (r_0, r_1)$, we deduce that $\lambda_0^A[a_{r_1}] = 0$. In consequence, from Theorem 11 and Lemma 25, we have that $L[a_{r_1}]$ satisfies the AMP.

Now, Corollary 8 gives us the following optimal estimate for the AMP:

• $L[a_r]$ satisfies the AMP on X_P if and only if $s \in (r_0, r_1]$.

In the previous example the estimation given by Theorem 17 to ensure the AMP is worse than the one obtained in Lemma 29. Now we present an example where Theorem 17 gives a better estimate for AMP.

Example 4. Consider the problem

$$u''(t) + \frac{\mu}{t \log^2 t} u(t) = h(t), \ t \in [0, 1/2], \quad u(0) = u(1/2), \quad u'(0) = u'(1/2),$$

with μ a positive constant. To study the values of the parameter $\mu > 0$ for which the MP or the AMP is ensured, we take into account that the potential $a_\mu(t) := \frac{\mu}{t \log^2 t}$ belongs to $L^1(0, 1/2)$, but it does not belong to $L^\alpha(0, 1/2)$ for any $\alpha > 1$.

Since $a_\mu(t) > 0$ for all $t \in (0, 1/2)$, we have, from Proposition 3, that the corresponding operator $L[a_\mu]$ cannot satisfy the MP.

On the other hand, from the fact that $\|(a_\mu)_+\|_1 = \|a_\mu\|_1 = \mu/\log(2)$ and $K(\infty, 1/2) = 8$, applying Lemma 29, we know that $L[a_\mu]$ satisfies the AMP for all $\mu \in (0, \mu_0]$, with $\mu_0 = 8 \log(2) \approx 5.54518$.

By means of Theorem 17, we can improve this estimation as follows: it is obvious that

$$\bar{a}_\mu = 2 \int_0^{1/2} a_\mu(s) \, ds = \frac{2\mu}{\log(2)},$$

and so we deduce that

$$\hat{a}_\mu(t) := a_\mu(t) - \bar{a}_\mu = \mu \left(\frac{1}{t \log^2 t} - \frac{2}{\log(2)} \right)$$

and

$$a_1 \equiv \left\| \left(\frac{1}{t \log^2 t} - \frac{2}{\log(2)} \right)_+ \right\|_1 \approx 0.26227.$$

Thus, condition imposed in Theorem 17 (3) is rewritten as

$$0 < \mu \le \frac{2\pi^2}{\frac{1}{\log(2)} + \frac{\pi^2}{4} a_1} \approx 9.44541,$$

which is a substantial improvement of the earlier estimate.

In this case it is not possible to get the explicit expressions of functions u_1 and u_2 and, as a consequence, of Δ_{a_μ}. However, we can study the related discrete equation and obtain this values with a very small error.

In particular, it is known (see [20]) that for a given $n \in \mathbb{N}$ large enough, the value $u_2(k/(2n)) \approx y(k)$, for all $k \in \{1, \dots, n\}$, where $y : \{0, \dots, n\} \to \mathbb{R}$ is the unique solution of the difference equation

$$y(k+1) - 2y(k) + y(k-1) + \frac{\mu}{2nk \log^2(k/(2n))} y(k) = 0, \quad k \in \{1, \dots, n-1\},$$

coupled with the initial conditions

$$y(0) = 0, \quad y(1) = 1/(2n).$$

In a similar way, we have an approximation of u_1, by considering the initial conditions $y(0) = y(1) = 1$.

So, by taking $n = 10^6$, we estimate the first root of the equation $\Delta_{a_\mu}(0) = -2$ by $\mu = \mu_1 \approx 11.6053$. Using Corollary 8 again, we deduce that the operator $L[a_\mu]$ satisfies the AMP on X_P if and only if $\mu \in (0, \mu_1]$.

Finally, we study the AMP for the Mathieu equation.

Example 5. In order to obtain 2π-periodic positive solutions for the Mathieu equation

$$\begin{aligned} u''(t) + (e + b\cos t)\, u(t) &= h(t), \quad t \in [0, 2\pi], \\ u(0) &= u(2\pi), \quad u'(0) = u'(2\pi), \end{aligned} \tag{3.17}$$

with $e \ge 0$, $b \in \mathbb{R}$, $e^2 + b^2 > 0$, and $h \succ 0$, an important tool is the validity of the AMP for operator

$$Lu := u'' + (e + b\cos t)\, u.$$

Notice that, since

$$\int_0^{2\pi} (e + b\cos s)\, ds = 2\pi\, e \ge 0,$$

only the AMP character of this equation has sense.

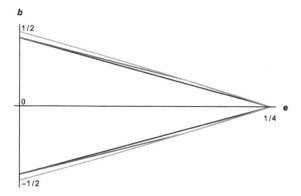

Figure 3.2 Admissible values of e and b to get the AMP for the Mathieu equation (wider set) together with the graphs of (3.19) (middle) and (3.18) (smaller).

Since we have that

$$\|(e \widehat{+ b \cos} t)_+\|_\alpha = |b| \, \pi^{\frac{1}{2\alpha}} \left(\frac{\Gamma\left(\frac{\alpha+1}{2}\right)}{\Gamma\left(\frac{\alpha+2}{2}\right)} \right)^{\frac{1}{\alpha}},$$

then Theorem 17, (3) means

$$0 \le |b| \le (1 - 4e) \max_{\alpha \ge 1} \left\{ \frac{K(2\alpha^*, 2\pi)}{\pi^{\frac{1}{2\alpha}} \left(\frac{\Gamma\left(\frac{\alpha+1}{2}\right)}{\Gamma\left(\frac{\alpha}{2}+1\right)} \right)^{\frac{1}{\alpha}}} \right\}, \tag{3.18}$$

which is shown in Fig. 3.2.

Moreover, as in the previous examples, Lemma 29 gives us an alternative estimation of the admissible values of e and b for which the Green's function is nonnegative, in this case

$$\|(e + b \cos t)_+\| \le \max_{\alpha \ge 1} \left\{ K(2\alpha^*, 2\pi) \right\}, \tag{3.19}$$

which is shown in Fig. 3.2 and, as we can observe, in this case is better than the one provided by Theorem 17 (3).

By using computational methods it is possible to approximate $\Delta_{e+b\cos t}(\lambda)$ (see Fig. 3.3). For instance, we have obtained:

- $\lambda_0^A[0] = 1/4$,
- $\lambda_0^A[(\cos t)/4)] \approx 0.17766$,

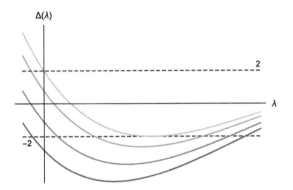

Figure 3.3 Discriminant related to (3.17) for $e = 0$ and $b = 0, 1/4, 2/5, 1/2$ (from above to below).

- $\lambda_0^A[2\,(\cos t)/5] \approx 0.031914$,
- $\lambda_0^A[(\cos t)/2] \approx -0.027562$.

Remark 20. The Brillouin-beam focusing equation

$$u''(t) + e\,(1 + \cos(t))\,u(t) = \frac{1}{u(t)},$$

appears in the study of electronics, and models the motion of a magnetically focused axially symmetric electron beam under the influence of a Brillouin flow (see [2] for details).

The existence of positive periodic solutions for this equation has been studied by several authors who gave some estimates on the parameter $e > 0$ to ensure such solutions (see for instance [15,16,27,29,30,33–35]). In some of the papers [29,30] the positiveness of the Green's function is fundamental to deduce the existence of solutions. Therefore the results obtained there are automatically valid for the range of parameters showed in Example 5.

However, in some cases the related Green's function changes sign at some of the values obtained by these authors [27]. So the positivity of the Green's function is a sufficient but not a necessary condition to ensure the existence of positive periodic solutions for the Brillouin equation.

3.4 NON-PERIODIC CONDITIONS

In this section we shall apply the results given in the previous one for the periodic problem, in order to assure the constant sign of the Green's function related to different types of boundary value problems. This way, we obtain conditions to warrant the existence of constant sign solutions for Neumann, Dirichlet, and mixed problems without doing a direct study of such problems. The key idea is that the expression of the Green's function related to each case can be obtained as a linear combination of the Green's function of the periodic problem. The results in this section are taken from [7].

From the expressions relating the different Green's functions, we are able to compare their constant sign. As a consequence, we deduce direct relations between the Green's functions. These results allow us to obtain comparison principles which warrant that, for certain intervals of the parameter λ, the solution of the nonhomogeneous Hill's equation under some suitable conditions is bigger in every point than the solution of the same equation under another type of boundary conditions.

We also obtain a decomposition of the spectrum of some problems as a combination of the other ones; this allows us to deduce a certain order of appearance for the eigenvalues of each problem. We include some numerical examples in which we observe an order of eigenvalues even more precise than the one theoretically proved.

The section finishes with a generalization of the results previously formulated in Subsections 3.3.3 and 3.3.4, which are reinterpreted in terms of non-periodic problems.

Through this section we will denote $J = [0, 2\,T]$. Moreover, when operator $L[a]$ is nonresonant we are going to denote its associated Green's function by $G[a, T]$ in order to stress its dependence on the length of the interval T. We will use an analogous notation, including the length of the interval, for eigenvalues.

We will study now different separated boundary conditions, and look for the connection between them and the periodic problem. In [1] some comparison principles were developed for these kind of boundary conditions. There it is proved that the validity of MP or AMP for one boundary condition is deduced from the validity of another one, considering for that some more restrictive hypothesis over the coefficients of the equation.

3.4.1 Neumann Problem

In this section we will obtain the expression of the Green's function of Neumann problem both as a sum of Green's functions of a periodic problem and as a sum of Green's functions of a Neumann problem defined in a different interval.

Assume now that the Neumann boundary value problem

$$L[a]\,u(t) = \sigma(t), \quad \text{a.e. } t \in I, \quad u'(0) = u'(T) = 0 \qquad (N,\,T)$$

has a unique solution $u \in W^{2,1}(I)$ for all $\sigma \in L^1(I)$.

Suppose, in addition, that the periodic boundary value problem

$$L[\tilde{a}]\,u(t) = \tilde{\sigma}(t), \quad \text{a.e. } t \in J, \quad u(0) = u(2\,T),\ u'(0) = u'(2\,T), \quad (P,\,2\,T)$$

with $\tilde{a}$ the even extension of a to the interval $[0, 2\,T]$, that is,

$$\tilde{a}(t) = \begin{cases} a(t), & \text{if } t \in [0,\,T], \\ a(2\,T - t), & \text{if } t \in [T, 2\,T], \end{cases}$$

has a unique solution $v \in W^{2,1}(J)$ for all $\tilde{\sigma} \in L^1(J)$.

Let u be the unique solution of problem $(N,\,T)$. It is clear that, by defining v as the even extension of u, we have that $v \in W^{2,1}(J)$ is a solution of $(P, 2\,T)$ for the particular case of $\tilde{\sigma}$ defined as the even extension of σ.

So, we have that for all $t \in I$ the following property holds:

$$\int_0^T G_N[a,\,T](t, s)\,\sigma(s)\,ds$$

$$= u(t) = v(t) = \int_0^{2\,T} G_P[\tilde{a}, 2\,T](t, s)\,\tilde{\sigma}(s)\,ds$$

$$= \int_0^T G_P[\tilde{a}, 2\,T](t, s)\,\sigma(s)\,ds$$

$$\quad + \int_T^{2\,T} G_P[\tilde{a}, 2\,T](t, s)\,\sigma(2\,T - s)\,ds$$

$$= \int_0^T \big(G_P[\tilde{a}, 2\,T](t, s) + G_P[\tilde{a}, 2\,T](t, 2\,T - s) \big)\,\sigma(s)\,ds.$$

Since $\sigma \in L^1(I)$ is arbitrarily chosen and due to the uniqueness of the Green's function of both problems, we arrive at the following connecting expression between Neumann and periodic Green's functions

$$G_N[a,\,T](t, s) = G_P[\tilde{a}, 2\,T](t, s) + G_P[\tilde{a}, 2\,T](t, 2\,T - s) \qquad \forall\, (t, s) \in I \times I.$$

$$(3.20)$$

Note that the previous equality gives us the exact expression of the Green's function for the Neumann problem by means of the value of the periodic one.

On the other hand, let $\tilde{\sigma} \in L^1(J)$ be arbitrarily chosen. Since the periodic problem $(P, 2\,T)$ has a unique solution v, we have, due to the fact that $\tilde{a}(t) = \tilde{a}(2\,T - t)$, that $w(t) = v(2\,T - t)$ is the unique solution of the periodic problem

$$L[\tilde{a}]\,w(t) = \tilde{\sigma}(2\,T - t), \quad \text{a.e. } t \in J, \quad w(0) = w(2\,T), \ w'(0) = w'(2\,T).$$

Since

$$v(2\,T - t) = \int_0^{2\,T} G_P[\tilde{a}, 2\,T](2\,T - t, s)\,\tilde{\sigma}(s)\,ds$$

and

$$w(t) = \int_0^{2\,T} G_P[\tilde{a}, 2\,T](t, s)\,\tilde{\sigma}(2\,T - s)\,ds = \int_0^{2\,T} G_P[\tilde{a}, 2\,T](t, 2\,T - s)\,\tilde{\sigma}(s)\,ds,$$

we arrive at

$$G_P[\tilde{a}, 2\,T](2\,T - t, s) = G_P[\tilde{a}, 2\,T](t, 2\,T - s) \qquad \forall\,(t, s) \in J \times J$$

or, which is the same,

$$G_P[\tilde{a}, 2\,T](t, s) = G_P[\tilde{a}, 2\,T](2\,T - t, 2\,T - s) \qquad \forall\,(t, s) \in J \times J.$$

In particular, we have that Eq. (3.20) can be rewritten as

$$G_N[a, T](t, s) = G_P[\tilde{a}, 2\,T](t, s) + G_P[\tilde{a}, 2\,T](2\,T - t, s) \qquad \forall\,(t, s) \in I \times I. \tag{3.21}$$

Moreover, we observe that v also satisfies Neumann boundary conditions on $[0, 2\,T]$ so it is a solution of the problem

$$u''(t) + \tilde{a}(t)\,u(t) = \tilde{\sigma}(t), \quad \text{a.e. } t \in J, \quad u'(0) = u'(2\,T) = 0. \qquad (N, 2\,T)$$

We will assume again that this solution is unique. An analogous reasoning lets us conclude that

$$G_N[a, T](t, s) = G_N[\tilde{a}, 2\,T](t, s) + G_N[\tilde{a}, 2\,T](2\,T - t, s) \qquad \forall\,(t, s) \in I \times I \tag{3.22}$$

or, using (3.21) and assuming that the periodic problem on the interval $[0, 4\,T]$ with potential $\tilde{\tilde{a}}$ (the even extension to $[0, 4\,T]$ of $\tilde{a}$) has also a unique solution,

$$
\begin{aligned}
G_N[a,\,T](t,\,s) =\; & G_P[\tilde{\tilde{a}},\,4\,T](t,\,s) + G_P[\tilde{\tilde{a}},\,4\,T](4\,T - t,\,s) \\
& + G_P[\tilde{\tilde{a}},\,4\,T](2\,T - t,\,s) \\
& + G_P[\tilde{\tilde{a}},\,4\,T](2\,T + t,\,s) \qquad \forall\,(t,\,s) \in I \times I.
\end{aligned}
$$

Note that $\tilde{\tilde{a}}$ is a $2\,T$-periodic function.

Observe that both a and $\tilde{a}$ or $\tilde{\tilde{a}}$ are not necessarily continuous functions.

3.4.2 Dirichlet Problem

In the same way, we will obtain the Green's function for the Dirichlet problem both as a sum of Green's functions for the periodic one and as a sum of Green's functions for another Dirichlet problem. Again, we will assume the uniqueness of solution for all the considered problems.

Let $G_D[a,\,T]$ be the Green's function related to the Dirichlet boundary value problem

$$
u''(t) + a(t)\,u(t) = \sigma(t), \qquad \text{a.e. } t \in I, \quad u(0) = u(T) = 0. \qquad (D,\,T)
$$

Making an odd extension v of function u to the interval J, we deduce that v is a solution of $(P,\,2\,T)$ for the particular choice of $\tilde{\sigma}$ as the odd extension of function σ to the interval J.

Reasoning as in the previous case, we conclude that

$$
G_D[a,\,T](t,\,s) = G_P[\tilde{a},\,2\,T](t,\,s) - G_P[\tilde{a},\,2\,T](2\,T - t,\,s) \qquad \forall\,(t,\,s) \in I \times I.
\tag{3.23}
$$

On the other hand, v also satisfies Dirichlet boundary conditions on $[0,\,2\,T]$ so it is a solution of the problem

$$
u''(t) + \tilde{a}(t)\,u(t) = \tilde{\sigma}(t), \qquad \text{a.e. } t \in J, \quad u(0) = u(2\,T) = 0, \qquad (D,\,2\,T)
$$

and we arrive at

$$
G_D[a,\,T](t,\,s) = G_D[\tilde{a},\,2\,T](t,\,s) - G_D[\tilde{a},\,2\,T](2\,T - t,\,s) \qquad \forall\,(t,\,s) \in I \times I
\tag{3.24}
$$

or, using (3.23),

$$G_D[a, T](t, s) = G_P[\tilde{\tilde{a}}, 4\,T](t, s) - G_P[\tilde{\tilde{a}}, 4\,T](4\,T - t, s)$$
$$- G_P[\tilde{\tilde{a}}, 4\,T](2\,T - t, s)$$
$$+ G_P[\tilde{\tilde{a}}, 4\,T](2\,T + t, s) \qquad \forall\,(t, s) \in I \times I.$$

3.4.3 Relation Between Neumann and Dirichlet Problems

From the expressions previously obtained for $G_N[a, T]$ and $G_D[a, T]$, we are going to connect the existence and uniqueness of solution, and consequently the spectrum, for Neumann, Dirichlet, and periodic problems.

As an immediate consequence of (3.21) and (3.23) we have

$$G_N[a, T](t, s) + G_D[a, T](t, s) = 2\,G_P[\tilde{a}, 2\,T](t, s) \qquad \forall\,(t, s) \in I \times I \tag{3.25}$$

and

$$G_N[a, T](t, s) - G_D[a, T](t, s) = 2\,G_P[\tilde{a}, 2\,T](2\,T - t, s) \qquad \forall\,(t, s) \in I \times I. \tag{3.26}$$

Remark 21. From equality (3.20) we have that if problem $(P, 2\,T)$ has a unique solution, then problem (N, T) has a solution given by

$$u(t) = \int_0^T \left(G_P[\tilde{a}, 2\,T](t, s) + G_P[\tilde{a}, 2\,T](2\,T - t, s) \right) \sigma(s)\,ds.$$

The uniqueness follows from the fact that the Neumann boundary conditions are linearly independent (see [4, Lemma 1.2.21]).

Consequently we observe that if problem $(P, 2\,T)$ is nonresonant, the same holds for (N, T). In other words, the sequence of eigenvalues of problem (N, T) is contained into the sequence of eigenvalues of $(P, 2\,T)$.

The same argument is valid, by means of equality (3.23), to ensure that if problem $(P, 2\,T)$ has a unique solution, then problem (D, T) has a unique solution too.

As a consequence, denoting by $\Lambda_N[a, T]$, $\Lambda_D[a, T]$, and $\Lambda_P[\tilde{a}, 2\,T]$ the corresponding set of eigenvalues of problems (N, T), (D, T), and $(P, 2\,T)$, we deduce that

$$\Lambda_N[a, T] \cup \Lambda_D[a, T] \subset \Lambda_P[\tilde{a}, 2\,T].$$

On the other hand, Eqs. (3.25) and (3.26) imply that the uniqueness of solution of problems (N, T) and (D, T) warrants the uniqueness of solution of $(P, 2T)$.

Thus, we conclude that

$$\Lambda_N[a, T] \cup \Lambda_D[a, T] = \Lambda_P[\tilde{a}, 2T].$$

The same reasoning lets us also deduce the following facts:
- If problem $(N, 2T)$ has a unique solution then (N, T) has a unique solution too.
- If problem $(D, 2T)$ has a unique solution then (D, T) has a unique solution too.
- The existence and uniqueness of solution of problem

$$\begin{aligned} &u''(t) + \tilde{\tilde{a}}(t)\, u(t) = \sigma(t), \quad \text{a.e. } t \in [0, 4T], \\ &u(0) = u(4T), \quad u'(0) = u'(4T), \end{aligned} \qquad (P, 4T)$$

is equivalent to the existence and uniqueness of solution of both $(N, 2T)$ and $(D, 2T)$ and, consequently, implies the existence and uniqueness of solution of (N, T) and (D, T).

As a consequence, denoting by $\Lambda_N[\tilde{a}, 2T]$, $\Lambda_D[\tilde{a}, 2T]$, and $\Lambda_P[\tilde{\tilde{a}}, 4T]$ the set of eigenvalues of $(N, 2T)$, $(D, 2T)$, and $(P, 4T)$, respectively, we have that

$$\begin{aligned} &\Lambda_N[a, T] \subset \Lambda_N[\tilde{a}, 2T], \\ &\Lambda_D[a, T] \subset \Lambda_D[\tilde{a}, 2T] \end{aligned}$$

and

$$\Lambda_N[a, T] \cup \Lambda_D[a, T] \subset \Lambda_N[\tilde{a}, 2T] \cup \Lambda_D[\tilde{a}, 2T] = \Lambda_P[\tilde{\tilde{a}}, 4T].$$

3.4.4 Mixed Problems and their Relation with Neumann and Dirichlet Ones

Consider now the mixed boundary value problem

$$u''(t) + a(t)\, u(t) = \sigma(t), \quad \text{a.e. } t \in I, \quad u'(0) = u(T) = 0. \qquad (M_1, T)$$

Making an odd extension v of function u to the interval J, we deduce that v is a solution of the antiperiodic problem

$$u''(t) + \tilde{a}(t)\,u(t) = \tilde{\sigma}(t), \quad \text{a.e. } t \in J,$$
$$u(0) = -u(2\,T), \quad u'(0) = -u'(2\,T), \qquad\qquad (A,\,2\,T)$$

with $\tilde{\sigma}$ the odd extension of σ to the interval J.

Thus we conclude that

$$G_{M_1}[a,\,T](t,\,s) = G_A[\tilde{a},\,2\,T](t,\,s) - G_A[\tilde{a},\,2\,T](2\,T - t,\,s) \qquad \forall\,(t,\,s) \in I \times I,$$
$$(3.27)$$

with $G_A[\tilde{a},\,2\,T]$ the Green's function related to $(A,\,2\,T)$.

But v also satisfies Neumann conditions on $[0,\,2\,T]$ so it is a solution of $(N,\,2\,T)$ and

$$G_{M_1}[a,\,T](t,\,s) = G_N[\tilde{a},\,2\,T](t,\,s) - G_N[\tilde{a},\,2\,T](2\,T - t,\,s) \qquad \forall\,(t,\,s) \in I \times I,$$
$$(3.28)$$

or, equivalently, using Eq. (3.21)

$$G_{M_1}[a,\,T](t,\,s) = G_P[\tilde{\tilde{a}},\,4\,T](t,\,s) + G_P[\tilde{\tilde{a}},\,4\,T](4\,T - t,\,s)$$
$$- G_P[\tilde{\tilde{a}},\,4\,T](2\,T - t,\,s)$$
$$- G_P[\tilde{\tilde{a}},\,4\,T](2\,T + t,\,s) \qquad \forall\,(t,\,s) \in I \times I.$$

We can study now the corresponding mixed boundary value problem

$$u''(t) + a(t)\,u(t) = \sigma(t), \quad \text{a.e. } t \in I, \quad u(0) = u'(T) = 0. \qquad (M_2,\,T)$$

Considering in this case v as the even extension of function u to the interval J, we conclude that v is a solution of problem $(A,\,2\,T)$ with $\tilde{\sigma}$ the even extension of σ. Reasoning as in previous cases, we deduce that

$$G_{M_2}[a,\,T](t,\,s) = G_A[\tilde{a},\,2\,T](t,\,s) + G_A[\tilde{a},\,2\,T](2\,T - t,\,s) \qquad \forall\,(t,\,s) \in I \times I.$$
$$(3.29)$$

In this case, v satisfies Dirichlet conditions on the interval $[0,\,2\,T]$, so it is also a solution of $(D,\,2\,T)$ and we have that

$$G_{M_2}[a,\,T](t,\,s) = G_D[\tilde{a},\,2\,T](t,\,s) + G_D[\tilde{a},\,2\,T](2\,T - t,\,s) \qquad \forall\,(t,\,s) \in I \times I,$$
$$(3.30)$$

or, which is the same, using Eq. (3.23)

$$
\begin{aligned}
G_{M_2}[a, T](t, s) = {} & G_P[\tilde{a}, 4\,T](t, s) - G_P[\tilde{a}, 4\,T](4\,T - t, s) \\
& + G_P[\tilde{a}, 4\,T](2\,T - t, s) \\
& - G_P[\tilde{a}, 4\,T](2\,T + t, s) \qquad \forall\, (t, s) \in I \times I.
\end{aligned}
$$

Consequently, from (3.27) and (3.29) we deduce the following properties

$$
G_{M_2}[a, T](t, s) + G_{M_1}[a, T](t, s) = 2\,G_A[\tilde{a}, 2\,T](t, s) \qquad \forall\, (t, s) \in I \times I
$$

and

$$
G_{M_2}[a, T](t, s) - G_{M_1}[a, T](t, s) = 2\,G_A[\tilde{a}, 2\,T](2\,T - t, s) \qquad \forall\, (t, s) \in I \times I.
$$

Analogously, from (3.22) and (3.28) we obtain that

$$
G_N[a, T](t, s) + G_{M_1}[a, T](t, s) = 2\,G_N[\tilde{a}, 2\,T](t, s) \qquad \forall\, (t, s) \in I \times I
\tag{3.31}
$$

and

$$
G_N[a, T](t, s) - G_{M_1}[a, T](t, s) = 2\,G_N[\tilde{a}, 2\,T](2\,T - t, s) \qquad \forall\, (t, s) \in I \times I
\tag{3.32}
$$

and, from (3.24) and (3.30) we deduce

$$
G_{M_2}[a, T](t, s) + G_D[a, T](t, s) = 2\,G_D[\tilde{a}, 2\,T](t, s) \qquad \forall\, (t, s) \in I \times I
\tag{3.33}
$$

and

$$
G_{M_2}[a, T](t, s) - G_D[a, T](t, s) = 2\,G_D[\tilde{a}, 2\,T](2\,T - t, s) \qquad \forall\, (t, s) \in I \times I.
\tag{3.34}
$$

Compiling all the previous equations we arrive at the following expression

$$
G_N[a, T] + G_D[a, T] + G_{M_1}[a, T] + G_{M_2}[a, T] = 4\,G_P[\tilde{a}, 4\,T] \quad \text{on } I \times I.
$$

Remark 22. Reasoning analogously to Remark 21 and taking into account the fact that all the considered boundary conditions are linearly independent, using [4, Lemma 1.2.21] we are able to deduce the following facts:

- The existence and uniqueness of solution of problems (M_1, T) and (M_2, T) is equivalent to the existence and uniqueness of solution of $(A, 2\,T)$.
- If problem $(P, 4\,T)$ has a unique solution, then (M_1, T) and (M_2, T) also have a unique solution and, consequently, problem $(A, 2\,T)$ has a unique solution too.
- The existence and uniqueness of solution of problems (N, T) and (M_1, T) is equivalent to the existence and uniqueness of solution of $(N, 2\,T)$.
- The existence and uniqueness of solution of problems (D, T) and (M_2, T) is equivalent to the existence and uniqueness of solution of $(D, 2\,T)$.
- From Remark 21, we know that the existence and uniqueness of solution of $(N, 2\,T)$ and $(D, 2\,T)$ is equivalent to the existence and uniqueness of solution of problem $(P, 4\,T)$. We deduce then that the periodic problem $(P, 4\,T)$ has a unique solution if and only if problems (N, T), (D, T), (M_1, T), and (M_2, T) have a unique solution.

As a consequence, denoting by $\Lambda_{M_1}[a, T]$, $\Lambda_{M_2}[a, T]$, and $\Lambda_A[\tilde{a}, 2\,T]$ the set of eigenvalues of problems (M_1, T), (M_2, T), and $(A, 2\,T)$, respectively, we conclude that

$$\Lambda_{M_1}[a, T] \cup \Lambda_{M_2}[a, T] = \Lambda_A[\tilde{a}, 2\,T] \subset \Lambda_P[\tilde{a}, 4\,T],$$
$$\Lambda_N[a, T] \cup \Lambda_{M_1}[a, T] = \Lambda_N[\tilde{a}, 2\,T],$$
$$\Lambda_D[a, T] \cup \Lambda_{M_2}[a, T] = \Lambda_D[\tilde{a}, 2\,T]$$

and

$$\Lambda_N[a, T] \cup \Lambda_D[a, T] \cup \Lambda_{M_1}[a, T] \cup \Lambda_{M_2}[a, T] = \Lambda_P[\tilde{a}, 4\,T].$$

Moreover, taking into account the previous inclusions and the ones obtained in Remark 21 it is clear that

$$\Lambda_P[\tilde{a}, 2\,T] = \Lambda_N[a, T] \cup \Lambda_D[a, T] \subset \Lambda_P[\tilde{a}, 4\,T],$$

and, consequently, we deduce

$$\Lambda_P[\tilde{a}, 2\,T] \cup \Lambda_A[\tilde{a}, 2\,T] = \Lambda_P[\tilde{\tilde{a}}, 4\,T].$$

On the other hand, it is possible to obtain a direct relation between the Green's functions of the two mixed problems.

Lemma 33. *Let $a \in L^{\alpha}(I)$ and define $b(t) = a(T - t)$ for all $t \in I$. Then*

$$G_{M_1}[a, T](T - t, T - s) = G_{M_2}[b, T](t, s) \qquad \forall \, (t, s) \in I \times I.$$

Proof. Let u be the unique solution for the mixed problem (M_1, T), given explicitly by

$$u(t) = \int_0^T G_{M_1}[a, T](t, s) \, \sigma(s) \, ds.$$

Clearly, $v(t) = u(T - t)$ is the unique solution of the mixed problem

$$v''(t) + b(t) \, v(t) = \sigma(T - t), \quad v(0) = v'(T) = 0.$$

Therefore, we know that

$$u(T - t) = \int_0^T G_{M_1}[a, T](T - t, s) \, \sigma(s) \, ds,$$

meanwhile, on the other hand, we have that

$$v(t) = \int_0^T G_{M_2}[b, T](t, s) \, \sigma(T - s) \, ds = \int_0^T G_{M_2}[b, T](t, T - s) \, \sigma(s) \, ds.$$

As the previous equalities are valid for all $\sigma \in L^1(I)$ we deduce that

$$G_{M_1}[a, T](T - t, s) = G_{M_2}[b, T](t, T - s) \qquad \forall \, (t, s) \in I \times I$$

or, equivalently,

$$G_{M_1}[a, T](T - t, T - s) = G_{M_2}[b, T](t, s) \qquad \forall \, (t, s) \in I \times I. \qquad \square$$

As an immediate consequence we have that

Corollary 10. *Let $a \in L^1(I)$ and define $b(t) = a(T - t)$ for all $t \in I$. Then*

$$\Lambda_{M_1}[a, T] = \Lambda_{M_2}[b, T].$$

In particular, if $a(t) = a(T - t)$ then the eigenvalues of both mixed problems are the same.

3.4.5 Order of Eigenvalues and Constant Sign of the Green's Function

This subsection starts with the adaptation of Proposition 2 for the Green's function of the Neumann problem. Next we prove the existence of a certain order of appearance for the first eigenvalues of each boundary value problem. Finally, we relate the constant sign of the diverse Green's functions considered in previous sections.

Let $\lambda_0^N[a, T]$ be the smallest eigenvalue of the Neumann problem

$$u''(t) + (a(t) + \lambda)\, u(t) = 0, \quad \text{a.e. } t \in I, \quad u'(0) = u'(T) = 0.$$

Let $\lambda_0^{M_1}[a, T]$ be the smallest eigenvalue of the mixed problem

$$u''(t) + (a(t) + \lambda)\, u(t) = 0, \quad \text{a.e. } t \in I, \quad u'(0) = u(T) = 0.$$

Let $\lambda_0^{M_2}[a, T]$ be the smallest eigenvalue of the mixed problem

$$u''(t) + (a(t) + \lambda)\, u(t) = 0, \quad \text{a.e. } t \in I, \quad u(0) = u'(T) = 0.$$

Lemma 34. *Suppose that the Green's function $G_N[a, T]$ is nonnegative on $I \times I$ and there is some $(t_0, s_0) \in I \times I$ for which $G_N[a, T](t_0, s_0) = 0$, then either $t_0 = s_0 = 0$ or $t_0 = s_0 = T$.*

Proof. Suppose that $G_N[a, T](t_0, s_0) = 0$ for some $(t_0, s_0) \in I \times I$. Since $G_N[a, T] \geq 0$ on $I \times I$, as operator $L[a]$ is self-adjoint, Proposition 2 lets us conclude that (t_0, s_0) belongs either to the boundary of the square of definition or to its diagonal.

In the first case, suppose that $t_0 \in (0, T)$ and $s_0 = 0$. Then we have that $x_0(t) \equiv G_N[a, T](t, 0)$ satisfies the equation

$$x_0''(t) + a(t)\, x_0(t) = 0, \ t \in (0, T], \quad x_0(t_0) = x_0'(t_0) = 0,$$

which means that $G_N[a, T](t, 0) \equiv 0$ on $(0, T]$.

From the symmetry of $G_N[a, T]$, we have that $G_N[a, T](0, s) \equiv 0$ for all $s \in (0, T]$.

As a consequence, $x_s(t) \equiv G_N[a, T](t, s)$ satisfies the equation

$$x_s''(t) + a(t)\, x_s(t) = 0, \ t \in [0, s), \quad x_s(0) = x_s'(0) = 0,$$

which implies that $G_N[a, T](t, s) \equiv 0$ for all $t < s$. Using again the symmetry of $G_N[a, T]$ we have that it is identically zero on $I \times I$ and we reach a contradiction.

The argument is valid for all (t_0, s_0) in the boundary of $I \times I$ except for $(0, 0)$ and (T, T).

Assume now that $G_N[a, T](t_0, t_0) = 0$ for some $t_0 \in (0, T)$. In this case, defining $x_{t_0}(t)$ as the even extension to J of $G_N[a, T](t, t_0)$, we have that it satisfies the equation

$$x_{t_0}''(t) + \tilde{a}(t) x_{t_0}(t) = 0, \ t \in (t_0, 2T - t_0), \quad x_{t_0}(t_0) = x_{t_0}(2T - t_0) = 0.$$

From Sturm's comparison theorem (Theorem 10), we have that for any $\lambda \geq 0$ every nontrivial solution of the equation

$$y''(t) + (\tilde{a}(t) + \lambda) y(t) = 0, \ t \in [0, 2T], \tag{3.35}$$

has as least one zero on $[t_0, 2T - t_0]$.

Then, as the even extension to J of the positive eigenfunction on $(0, T]$ associated with $\lambda_0^{M_2}[a, T]$ solves (3.35), we deduce that $\lambda_0^{M_2}[a, T] < 0$.

As a consequence, for any $\lambda \in (\lambda_0^{M_2}[a, T], 0]$ we have that y_0, the even extension to J of $G_N[a + \lambda, T](t, 0)$, has at least one zero on $(0, 2T)$. Moreover, all the zeros of y_0 are simple because otherwise $G_N[a + \lambda, T](t, 0) \equiv 0$ on $(0, T]$, which cannot happen. Then necessarily y_0 changes its sign on $(0, 2T)$ and, as it is an even function, $G_N[a + \lambda, T](t, 0)$ changes its sign on $(0, T)$. This contradicts the hypothesis that $G_N[a, T]$ is nonnegative on $I \times I$.   □

Note that if $G_N[a, T](0, 0) = 0$ we have that $x(t) \equiv G_N[a, T](t, 0)$ is a solution of

$$x''(t) + a(t) x(t) = 0, \ t \in I, \quad x(0) = x'(T) = 0. \tag{3.36}$$

Moreover, when $G_N[a, T](T, T) = 0$, $y(t) = G_N[a, T](t, T)$ is a solution of

$$y''(t) + a(t) y(t) = 0, \ t \in I, \quad y'(0) = y(T) = 0. \tag{3.37}$$

As a consequence of previous result and equality (3.20), we deduce the following result.

Corollary 11. *If $G_P[\tilde{a}, 2T]$ has constant sign on $J \times J$ then $G_N[a, T]$ has the same sign as $G_P[\tilde{a}, 2T]$ on $I \times I$ and it is different from zero for all $(t, s) \in (I \times I) \setminus \{(0, 0) \cup (T, T)\}$.*

Moreover, $G_N[a, T](T, T) = 0$ if and only if Eq. (3.36) has a nonzero and constant sign solution on $[0, T)$, which means that $\lambda_0^{M_2}[a, T] = 0$.

$G_N[a, T](0, 0) = 0$ if and only if Eq. (3.37) has a nonzero and constant sign solution on $(0, T]$, which means that $\lambda_0^{M_1}[a, T] = 0$.

Reasoning as in Lemma 34, it is proved the following.

Lemma 35. *Suppose that $a \in L^1(I)$, the Green's function $G_D[a, T]$ has constant sign on $I \times I$ and there exists some $(t_0, s_0) \in I \times I$ such that $G_D[a, T](t_0, s_0) = 0$. Then (t_0, s_0) belongs to the boundary of the square of definition of $G_D[a, T]$.*

Remark 23. From the Dirichlet boundary conditions and property (G5) of the Green's function, it is clear that $G_D[a, T]$ must cancel on the whole boundary of $I \times I$. Previous lemma ensures that, when $G_D[a, T]$ has constant sign, it can not vanish at any other point.

Moreover, it can be deduced the following.

Lemma 36. *Suppose that $a \in L^1(I)$, then:*
$G_D[a + \lambda, T] < 0$ on $(0, T) \times (0, T)$ if and only if $\lambda < \lambda_0^D[a, T]$.
Moreover, if $\lambda > \lambda_0^D[a, T]$ is such that $G_D[a + \lambda, T]$ exists, then $G_D[a + \lambda, T]$ changes sign.

Proof. Choose $\lambda < \lambda_0^D[a, T]$. From Theorem 10 it is clear that any solution of equation

$$u''(t) + (a(t) + \lambda)\, u(t) = 0, \quad t \in I, \tag{3.38}$$

has at most one zero on I.

As we have seen in Definition 10, for each fixed $s_0 \in (0, T)$, $u_{s_0}(\cdot) := G_D[a + \lambda, T](\cdot, s_0)$ satisfies (3.38) on $[0, s_0) \cup (s_0, T]$.

Then, if u is the unique solution of (3.38) under the initial conditions

$$u(0) = 0, \quad u'(0) = 1,$$

it is clear that there exists a nonzero constant k_1 such that $u_{s_0}(t) = k_1\, u(t)$ for all $t < s_0$. Therefore,

$$G_D[a + \lambda, T](t, s) = k_1(s)\, u(t) \quad \text{for all } t < s.$$

Moreover, since $u(0) = 0$, we have that $u(t) \neq 0$ for all $t \in (0, T]$.

Analogously, if v is the unique solution of (3.38) satisfying the final conditions

$$u(T) = 0, \quad u'(T) = -1,$$

then there exists a nonzero constant k_2 such that $u_{s_0}(t) = k_2\, v(t)$ for all $t > s_0$. Consequently,

$$G_D[a + \lambda, T](t, s) = k_2(s)\, v(t) \quad \text{for all } s < t.$$

In this case, $v(T) = 0$ implies that $v(t) \neq 0$ for all $t \in [0, T)$.

Now, since $G_D[a + \lambda, T]$ is a symmetric function, necessarily $k_1(s) = c\, v(s)$ and $k_2(s) = c\, u(s)$ for some nonzero constant c, that is,

$$G_D[a + \lambda, T](t, s) = \begin{cases} c\, v(s)\, u(t), & 0 \leq t < s \leq T, \\ c\, u(s)\, v(t), & 0 \leq s < t \leq T, \end{cases}$$

and, since $G_D[a + \lambda, T]$ is continuous on $I \times I$, it is clear that

$$G_D[a + \lambda, T](s, s) = c\, u(s)\, v(s).$$

Therefore $G_D[a + \lambda, T]$ has strict constant sign on $(0, T) \times (0, T)$ for all $\lambda < \lambda_0^D[a, T]$.

From this property, since $\frac{\partial\, G[a+\lambda, T]}{\partial t}(0, s) \neq 0$ and $\frac{\partial\, G[a+\lambda, T]}{\partial t}(T, s) \neq 0$ for all $s \in (0, T)$, it is immediate to verify that choosing $\phi(t) = t(T - t)$, for all $s \in (0, T)$ we have that

$$k_1(s) = \min_{t \in I} \frac{|G_D[a + \lambda, T](t, s)|}{\phi(t)} \in (0, \infty)$$

and

$$k_2(s) = \max_{t \in I} \frac{|G_D[a + \lambda, T](t, s)|}{\phi(t)} \in (0, \infty)$$

and are continuous functions on I.

So, if there exists some $\bar\lambda < \lambda_0^D[a, T]$ for which $G_D[a + \bar\lambda, T] > 0$ on $(0, T) \times (0, T)$ then property (P_g) is fulfilled and, from Lemma 16, taking into account that

$$\lambda_0^D[a + \bar\lambda, T] = \lambda_0^D[a, T] - \bar\lambda,$$

we have that a necessary condition for $G_D[a + \bar\lambda + \mu, T]$ to be nonnegative on $I \times I$ is that $\mu > \lambda_0^D[a, T] - \bar\lambda$ or, which is the same, if $G_D[a + \lambda, T] \geq 0$ on $I \times I$ then $\lambda > \lambda_0^D[a, T]$. This facts contradicts the existence of such $\bar\lambda$.

As a consequence, $G_D[a + \lambda, T] < 0$ on $(0, T) \times (0, T)$ for all $\lambda < \lambda_0^D[a, T]$ and condition (N_g) is fulfilled.

Thus, from Lemma 15, we can ensure that $G_D[a + \lambda, T] < 0$ on $(0, T) \times (0, T)$ if and only if $\lambda < \lambda_0^D[a, T]$.

Now we will see that for $\lambda > \lambda_0^D[a, T]$ such that the Green's function $G_D[a + \lambda, T]$ exists, it holds that $G_D[a + \lambda, T]$ changes sign.

Take $\lambda_0^D[a, T] < \lambda < \lambda_1^D[a, T]$. In Chapter 2 we have proved that the eigenfunction related to $\lambda_0^D[a, T]$ cancels on 0 and T but has strict constant sign on $(0, T)$, and that the eigenfunction related to $\lambda_1^D[a, T]$ cancels on 0 and T and has exactly one more zero on $(0, T)$. Taking this into account and using Theorem 10, it is easy to deduce that any solution of equation

$$u''(t) + (a(t) + \lambda) u(t) = 0, \ t \in I$$

has exactly one zero on $(0, T)$.

In particular, both u and v, defined in the first part of this proof, have exactly one zero on $(0, T)$. Let's denote by t_0 and t_1 the zeros of u and v, respectively, and assume that $t_1 < t_0$ (being the other case analogous).

Now, for $s < t_1$, $u(s) > 0$ and $v(s) < 0$, meanwhile for $t_1 < r < t_0$, $u(r) > 0$ and $v(r) > 0$. This way we have found two points $(s, s), (r, r) \in (0, T) \times (0, T)$ such that $G_D[a + \lambda, T](s, s) \, G_D[a + \lambda, T](r, r) < 0$.

As a consequence, for $\lambda_0^D[a, T] < \lambda < \lambda_1^D[a, T]$, $G_D[a + \lambda, T]$ changes sign and so from Lemma 16 we conclude that $G_D[a + \lambda, T]$ must change sign for every $\lambda > \lambda_0^D[a, T]$. □

Remark 24. Previous result is a particular case ($n = 2$, $k = 1$) of [11, Theorem 3.1], where a characterization of the constant sign of the Green's function is proved for the general n-th-order linear operator coupled with the so-called $(k, n - k)$ boundary conditions.

It can be also deduced as a consequence of [13, Theorem 11], where it is showed that if a linear equation is disconjugate then the related Green's function has constant sign, and [12, Theorem 2.1], where the interval of disconjugation is characterized by means of the eigenvalues of some suitable boundary condition.

Remark 25. We note that the fact that the function

$$\lambda_0^D : L^\alpha(I) \to \mathbb{R}$$
$$a \mapsto \lambda_0^D[a]$$

is decreasing (which is a particular case of Lemma 2) could also be deduced by arguing as in Lemma 28, taking into account the previous lemma and the fact that the eigenfunction related to the first eigenvalue of the Dirichlet problem has constant sign on $(0, T)$.

In the sequel we prove some relations between the first eigenvalues of various problems. To do that, we will take into account that, in view of Lemma 10, it is possible to rewrite Lemma 25 in terms of the sign of the Green's function.

Lemma 37. *Suppose that $a \in L^{\alpha}(I)$, then:*
1. $G_P[a + \lambda, T] < 0$ *if and only if* $\lambda < \lambda_0^P[a, T]$.
2. $G_P[a + \lambda, T] \geq 0$ *if and only if* $\lambda_0^P[a, T] < \lambda \leq \lambda_0^A[a, T]$.

Then, we are in a position to prove the following result.

Theorem 18. *The following equalities are fulfilled for any $a \in L^{\alpha}(I)$.*
1. $\lambda_0^N[a, T] = \lambda_0^N[\tilde{a}, 2\,T] = \lambda_0^P[\tilde{a}, 2\,T]$.
2. $G_N[a + \lambda, T](t, s) < 0$ *for all* $(t, s) \in I \times I$ *if and only if* $\lambda < \lambda_0^P[\tilde{a}, 2\,T]$.
3. $G_N[a + \lambda, T](t, s) > 0$ *on* $I \times I$ *if and only if*

$$\lambda_0^P[\tilde{a}, 2\,T] < \lambda < \min\{\lambda_0^{M_1}[a, T], \lambda_0^{M_2}[a, T]\}.$$

4. $G_N[a + \lambda, T](t, s) > 0$ *for all* $(t, s) \in (I \times I)\backslash\{(0, 0) \cup (T, T)\}$ *if and only if*

$$\lambda_0^P[\tilde{a}, 2\,T] < \lambda \leq \min\{\lambda_0^{M_1}[a, T], \lambda_0^{M_2}[a, T]\}.$$

5. $G_N[a, T](0, 0) = 2\,G_P[\tilde{a}, 2\,T](0, 0)$ *and* $\lambda_0^{M_1}[a, T]$ *is characterized as the first root of equation*

$$(G_N[a + \lambda, T](0, 0) =)\ G_P[\tilde{a} + \lambda, 2\,T](0, 0) = 0.$$

6. $G_N[a, T](T, T) = 2\,G_P[\tilde{a}, 2\,T](T, T)$ *and* $\lambda_0^{M_2}[a, T]$ *is characterized as the first root of equation*

$$(G_N[a + \lambda, T](T, T) =)\ G_P[\tilde{a} + \lambda, 2\,T](T, T) = 0.$$

7. $\lambda_0^A[\tilde{a}, 2\,T] = \min\{\lambda_0^{M_1}[a, T], \lambda_0^{M_2}[a, T]\}$.
8. $\lambda_0^{M_2}[a, T] = \lambda_0^D[\tilde{a}, 2\,T] < \lambda_0^D[a, T]$.
9. $\lambda_0^N[a, T] < \lambda_0^{M_1}[a, T]$.

Proof. Suppose that the periodic problem $(P, 2\,T)$ is uniquely solvable. From Remark 21 we know that the Neumann problem (N, T) is uniquely solvable too.

From Lemmas 19 and 25, we know that $G_P[\tilde{a} + \lambda, 2\,T]$ is strictly negative on $J \times J$ if and only if $\lambda < \lambda_0^P[\tilde{a}, 2\,T]$ and it is nonnegative on $J \times J$ if and only if $\lambda_0^P[\tilde{a}, 2\,T] < \lambda \leq \lambda_0^A[\tilde{a}, 2\,T]$.

As a consequence, Eq. (3.21) implies that $G_N[a+\lambda, T]$ is strictly negative on $I \times I$ (and consequently condition (N_g) introduced in Lemma 15 is verified) for all $\lambda < \lambda_0^P[\tilde{a}, 2\,T]$ and it is nonnegative on $I \times I$ (and satisfies (P_g) in Lemma 16) for all $\lambda_0^P[\tilde{a}, 2\,T] < \lambda \le \lambda_0^A[\tilde{a}, 2\,T]$.

Since, as it is pointed out in Lemma 15, the set of parameters λ for which the maximum principle holds is connected, and its supremum is the first eigenvalue of the considered operator, we conclude that $G_N[a+\lambda, T]$ is strictly negative on $I \times I$ if and only if $\lambda < \lambda_0^P[\tilde{a}, 2\,T]$ and $\lambda_0^N[a, T] = \lambda_0^P[\tilde{a}, 2\,T]$.

Let u_N be an eigenfunction associated to $\lambda_0^N[a, T]$. As we have seen in Chapter 2, we know that $\lambda_0^N[a, T]$ is simple and u_N is strictly positive on $(0, T)$. In fact it is strictly positive on the closed interval because, on the contrary, we would have that $u_N(0) = u_N'(0) = 0$ (or $u_N(T) = u_N'(T) = 0$) and we would conclude that u_N is identically zero on I.

Considering now the even extension of u_N to the interval J, we have that it remains an eigenfunction associated to the same value $\lambda_0^N[a, T]$ for the potential $\tilde{a}$ defined on the interval J. Since it is strictly positive on J, we deduce that $\lambda_0^N[a, T]$ is the smallest eigenvalue of $(N, 2\,T)$, that is, $\lambda_0^N[a, T] = \lambda_0^N[\tilde{a}, 2\,T]$.

So the two first identities are proved.

On the other hand, we know that $G_N[a+\lambda, T]$ is nonnegative on $I \times I$ for all $\lambda_0^P[\tilde{a}, 2\,T] < \lambda \le \lambda_0^A[\tilde{a}, 2\,T]$. So Lemma 16 implies that the maximum λ for which the Green's function is nonnegative on $I \times I$ is not an eigenvalue of the Neumann problem. Lemma 34 and Corollary 11 ensure that such value is exactly $\min\{\lambda_0^{M_1}[a, T], \lambda_0^{M_2}[a, T]\}$. So assertions three and four are proved.

Now, taking into account equality (3.21) and the fact that

$$G_P[\tilde{a}+\lambda, 2\,T](0, 0) = G_P[\tilde{a}+\lambda, 2\,T](2\,T, 0),$$

we conclude that for all $\lambda \in \mathbb{R}$ the following equalities hold

$$G_N[a+\lambda, T](0, 0) = 2\,G_P[\tilde{a}+\lambda, 2\,T](0, 0) \tag{3.39}$$

and

$$G_N[a+\lambda, T](T, T) = 2\,G_P[\tilde{a}+\lambda, 2\,T](T, T). \tag{3.40}$$

From Lemma 23, we have that while both values on Eqs. (3.39) and (3.40) are positive, they are strictly decreasing with respect to λ. Thus,

Corollary 11 ensures that $\lambda_0^{M_1}[a, T]$ is the first zero of (3.39) and $\lambda_0^{M_2}[a, T]$ the first zero of (3.40). Then, assertions five and six hold.

Assertion seven is an immediate consequence of

$$\Lambda_A[\tilde{a}, 2\,T] = \Lambda_{M_1}[a, T] \cup \Lambda_{M_2}[a, T].$$

To see assertion eight, it is enough to consider the even extension to J of the eigenfunction associated to $\lambda_0^{M_2}[a, T]$. Obviously it satisfies Dirichlet boundary conditions and is positive on $(0, 2\,T)$. Thus, $\lambda_0^{M_2}[a, T] = \lambda_0^D[\tilde{a}, 2\,T]$.

Taking into account that if we consider the odd extension to J of the eigenfunction related to the Dirichlet problem with $\lambda_0^D[a, T]$ (which is positive on $(0, T)$), we obtain a sign-changing eigenfunction of $(D, 2\,T)$. Consequently, $\lambda_0^D[a, T]$ is an eigenvalue of this problem too, but it is not the least one because the associated eigenfunction has not constant sign on $(0, 2\,T)$. As a consequence, $\lambda_0^D[\tilde{a}, 2\,T] < \lambda_0^D[a, T]$.

The same reasoning is valid to prove assertion nine. Indeed, the odd extension to J of the eigenfunction related to $\lambda_0^{M_1}[a, T]$ (which is positive on $(0, T)$) is an eigenfunction of $(N, 2\,T)$ and changes its sign on J. Consequently $\lambda_0^{M_1}[a, T] > \lambda_0^N[\tilde{a}, 2\,T] = \lambda_0^N[a, T]$.   □

Remark 26. Corollary 10 assures that both $\lambda_0^A[\tilde{a}, 2\,T] = \lambda_0^{M_1}[a, T]$ and $\lambda_0^A[\tilde{a}, 2\,T] = \lambda_0^{M_2}[a, T]$ are possible in assertion four of the previous theorem. Indeed, if for some potential a, $\lambda_0^{M_1}[a, T] < \lambda_0^{M_2}[a, T]$, then necessarily $\lambda_0^{M_2}[b, T] < \lambda_0^{M_1}[b, T]$ for $b(t) = a(T - t)$.

For an arbitrary potential a we obtain the following corollaries.

Corollary 12. *The following equalities are fulfilled for any $a \in L^\alpha(I)$.*

1. $G_N[a, T](t, s) < 0$ *for all $(t, s) \in I \times I$ if and only if $0 < \lambda_0^P[\tilde{a}, 2\,T]$* $(= \lambda_0^N[a, T])$.

2. $G_N[a, T](t, s) > 0$ *for all $(t, s) \in (0, T) \times (0, T)$ if and only if*

$$(\lambda_0^N[a, T] =) \lambda_0^P[\tilde{a}, 2\,T] < 0 \le \min\{\lambda_0^{M_1}[a, T], \lambda_0^{M_2}[a, T]\}$$
$$(\le \lambda_0^D[\tilde{a}, 2\,T] < \lambda_0^D[a, T]).$$

Corollary 13. *The following property holds for any $a \in L^\alpha(I)$:*
    $G_{M_2}[a + \lambda] < 0$ *on $(0, T] \times (0, T]$ if and only if $\lambda < \lambda_0^{M_2}[a, T]$.*

Proof. From Lemma 36 we know that $G_D[\tilde{a} + \lambda, 2\,T]$ is strictly negative on $(0, 2\,T) \times (0, 2\,T)$ if and only if $\lambda < \lambda_0^D[\tilde{a}, 2\,T]$.

Therefore, considering Eq. (3.30), it is immediately deduced that when $\lambda < \lambda_0^D[\tilde{a}, 2\,T]$, $G_{M_2}[a, T]$ is strictly negative on $(0, T] \times (0, T]$. We conclude the result from Lemma 15 and the fact that $\lambda_0^D[\tilde{a}, 2\,T] = \lambda_0^{M_2}[a, T]$. $\square$

Moreover, as an immediate consequence of Corollaries 10 and 13 we have

Corollary 14. *Let* $a \in L^\alpha(I)$, *then:*
$G_{M_1}[a + \lambda] < 0$ *on* $[0, T) \times [0, T)$ *if and only if* $\lambda < \lambda_0^{M_1}[a, T]$.

Remark 27. We note that the fact that the functions

$$\lambda_0^N : L^\alpha(I) \to \mathbb{R}$$
$$a \mapsto \lambda_0^N[a],$$
$$\lambda_0^{M_1} : L^\alpha(I) \to \mathbb{R}$$
$$a \mapsto \lambda_0^{M_1}[a]$$

and

$$\lambda_0^{M_2} : L^\alpha(I) \to \mathbb{R}$$
$$a \mapsto \lambda_0^{M_2}[a]$$

are decreasing (which are particular cases of Lemmas 3, 4, and 5) could also be deduced by arguing as in Lemma 28, taking into account the previous corollaries and the fact that the eigenfunctions related to the first eigenvalues of the Neumann and mixed problems have constant sign.

From the previous results we deduce some relations between the constant sign of the Green's functions.

Corollary 15. *For any* $a \in L^1(I)$ *we have the following properties.*
1. $G_P[\tilde{a}, 2\,T] < 0$ *on* $J \times J$ *if and only if* $G_N[a, T] < 0$ *on* $I \times I$. *This is equivalent to* $G_N[\tilde{a}, 2\,T] < 0$ *on* $J \times J$.
2. $G_P[\tilde{a}, 2\,T] > 0$ *on* $(0, 2\,T) \times (0, 2\,T)$ *if and only if* $G_N[a, T] > 0$ *on* $(0, T) \times (0, T)$.
3. *If* $G_N[\tilde{a}, 2\,T] > 0$ *on* $(0, 2\,T) \times (0, 2\,T)$ *then* $G_N[a, T] > 0$ *on* $(0, T) \times (0, T)$.
4. *If* $G_P[\tilde{a}, 2\,T] < 0$ *on* $J \times J$ *then* $G_D[\tilde{a}, 2\,T] < 0$ *on* $(0, 2\,T) \times (0, 2\,T)$.
5. *If* $G_P[\tilde{a}, 2\,T] > 0$ *on* $(0, 2\,T) \times (0, 2\,T)$ *then* $G_D[\tilde{a}, 2\,T] < 0$ *on* $(0, 2\,T) \times (0, 2\,T)$.

6. If $G_N[a, T]$ (or $G_P[\tilde{a}, 2\,T]$) has constant sign on $I \times I$, then $G_D[a, T] < 0$ on $(0, T) \times (0, T)$, $G_{M_1}[a, T] < 0$ on $[0, T) \times [0, T)$, and $G_{M_2}[a, T] < 0$ on $(0, T] \times (0, T]$.

7. $G_D[\tilde{a}, 2\,T] < 0$ on $(0, 2\,T) \times (0, 2\,T)$ if and only if $G_{M_2}[a, T] < 0$ on $(0, T] \times (0, T]$.

8. If either $G_{M_2}[a, T] < 0$ on $(0, T] \times (0, T]$ or $G_{M_1}[a, T] < 0$ on $[0, T) \times [0, T)$, then $G_D[a, T] < 0$ on $(0, T) \times (0, T)$.

3.4.6 Relations Between Green's Functions. Comparison Principles

From the relations between the constant sign of the Green's functions (Corollary 15), as well as the explicit expressions obtained in previous subsections for such functions, we will deduce some comparison criteria for Green's function of different problems and, as a consequence, for solutions of the related nonhomogeneous linear problems.

From (3.25) and (3.26), we obtain the following result.

Corollary 16. If $G_P[\tilde{a}, 2\,T] \geq 0$ on $J \times J$, then

$$G_N[a, T](t, s) \geq |G_D[a, T](t, s)| \, (= -G_D[a, T](t, s)) \qquad \forall\, (t, s) \in I \times I.$$

If $G_P[\tilde{a}, 2\,T] < 0$ on $J \times J$, then

$$G_N[a, T](t, s) < G_D[a, T](t, s) \, (\leq 0) \qquad \forall\, (t, s) \in I \times I.$$

As a consequence, we deduce the following comparison principles.

Theorem 19. Suppose that $G_P[\tilde{a}, 2\,T] \geq 0$ on $J \times J$. Let u_N be the unique solution of problem (N, T) for $\sigma = \sigma_1$ and u_D the unique solution of problem (D, T) for $\sigma = \sigma_2$.

Suppose that $|\sigma_2(t)| \leq \sigma_1(t)$ a.e. $t \in I$, then $|u_D(t)| \leq u_N(t)$ for all $t \in I$.

Proof. By definition of Green's functions, as a consequence of Corollary 16 we have the following inequalities for every $t \in I$:

$$
\begin{aligned}
|u_D(t)| &= \left| \int_0^T G_D[a, T](t, s)\, \sigma_2(s)\, ds \right| \leq \int_0^T |G_D[a, T](t, s)|\, |\sigma_2(s)|\, ds \\
&\leq \int_0^T G_N[a, T](t, s)\, |\sigma_2(s)|\, ds \leq \int_0^T G_N[a, T](t, s)\, \sigma_1(s)\, ds = u_N(t).
\end{aligned}
$$

$\square$

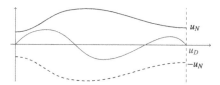

Figure 3.4 Solutions of (N, T) and (D, T) in conditions of Theorem 19.

Figure 3.5 Solutions of (N, T) and (D, T) in Case 1 in Theorem 20.

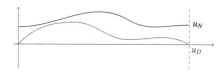

Figure 3.6 Solutions of (N, T) and (D, T) in Case 2 in Theorem 20.

The situation described in previous theorem is represented in Fig. 3.4.

Theorem 20. *Suppose that $G_P[\tilde{a}, 2\,T] < 0$ on $J \times J$. Let u_D be the unique solution of problem (D, T) for $\sigma = \sigma_1$ and u_N the unique solution of problem (N, T) for $\sigma = \sigma_2$.*

1. *If $0 \leq \sigma_2(t) \leq \sigma_1(t)$ a.e. $t \in I$, then $u_N(t) \leq u_D(t) \leq 0$ for all $t \in I$.*
2. *If $0 \geq \sigma_2(t) \geq \sigma_1(t)$ a.e. $t \in I$, then $u_N(t) \geq u_D(t) \geq 0$ for all $t \in I$.*

The graphical representation of these two situations can be seen in Figs. 3.5 and 3.6.

Analogously, from (3.31), (3.32), (3.33), and (3.34) we conclude the following.

Corollary 17. *If $G_N[\tilde{a}, 2\,T] \geq 0$ on $J \times J$, then*

$$G_N[a, T](t, s) \geq |G_{M_1}[a, T](t, s)| \, (= -G_{M_1}[a, T](t, s)) \qquad \forall \, (t, s) \in I \times I.$$

If $G_N[\tilde{a}, 2\,T] < 0$ on $J \times J$, then

$$G_N[a, T](t, s) < G_{M_1}[a, T](t, s) \, (\leq 0) \qquad \forall \, (t, s) \in I \times I.$$

If $G_D[\tilde{a}, 2\,T] \le 0$ on $J \times J$, then

$$G_{M_2}[a, T](t, s) < G_D[a, T](t, s) \,(\le 0) \qquad \forall\,(t, s) \in I \times I.$$

As a consequence, we deduce the following comparison principles.

Theorem 21. *Suppose that $G_N[\tilde{a}, 2\,T] \ge 0$ on $J \times J$. Let u_N be the unique solution of problem $(N,\ T)$ for $\sigma = \sigma_1$ and u_{M_1} the unique solution of $(M_1,\ T)$ for $\sigma = \sigma_2$.*

Suppose that $|\sigma_2(t)| \le \sigma_1(t)$ a.e. $t \in I$, then $|u_{M_1}(t)| \le u_N(t)$ for all $t \in I$.

Theorem 22. *Suppose that $G_N[\tilde{a}, 2\,T] < 0$ on $J \times J$. Let u_{M_1} be the unique solution of problem $(M_1,\ T)$ for $\sigma = \sigma_1$ and u_N the unique solution of $(N,\ T)$ for $\sigma = \sigma_2$.*

1. *If $0 \le \sigma_2(t) \le \sigma_1(t)$ a.e. $t \in I$, then $u_N(t) \le u_{M_1}(t) \le 0$ for all $t \in I$.*
2. *If $0 \ge \sigma_2(t) \ge \sigma_1(t)$ a.e. $t \in I$, then $u_N(t) \ge u_{M_1}(t) \ge 0$ for all $t \in I$.*

Theorem 23. *Suppose that $G_D[\tilde{a}, 2\,T] \le 0$ on $J \times J$. Let u_D be the unique solution of problem $(D,\ T)$ for $\sigma = \sigma_1$ and u_{M_2} the unique solution of problem $(M_2,\ T)$ for $\sigma = \sigma_2$.*

1. *If $0 \le \sigma_2(t) \le \sigma_1(t)$ a.e. $t \in I$, then $u_{M_2}(t) \le u_D(t) \le 0$ for all $t \in I$.*
2. *If $0 \ge \sigma_2(t) \ge \sigma_1(t)$ a.e. $t \in I$, then $u_{M_2}(t) \ge u_D(t) \ge 0$ for all $t \in I$.*

Rewriting Corollaries 16 and 17 in terms of eigenvalues, we have the following result.

Corollary 18. *If $(\lambda_0^N[a, T] =) \lambda_0^P[\tilde{a}, 2\,T] < 0 \le \lambda_0^A[\tilde{a}, 2\,T]$, then*

$$G_N[a, T](t, s) \ge -G_D[a, T](t, s) \ge 0 \qquad \forall\,(t, s) \in I \times I.$$

If $(\lambda_0^N[a, T] = \lambda_0^N[\tilde{a}, 2\,T] = \lambda_0^P[\tilde{\tilde{a}}, 4\,T] =) \lambda_0^P[\tilde{a}, 2\,T] > 0$, then

$$G_N[a, T](t, s) < G_D[a, T](t, s) \le 0 \qquad \forall\,(t, s) \in I \times I$$

and

$$G_N[a, T](t, s) < G_{M_1}[a, T](t, s) \le 0 \qquad \forall\,(t, s) \in I \times I.$$

If $(\lambda_0^N[a, T] = \lambda_0^N[\tilde{a}, 2\,T] =) \lambda_0^P[\tilde{\tilde{a}}, 4\,T] < 0 \le \lambda_0^A[\tilde{\tilde{a}}, 4\,T]$, then

$$G_N[a, T](t, s) \ge -G_{M_1}[a, T](t, s) \ge 0 \qquad \forall\,(t, s) \in I \times I.$$

If $(\lambda_0^D[\tilde{a}, 2\,T] =) \lambda_0^{M_2}[a, T] > 0$, then

$$G_{M_2}[a, T](t, s) < G_D[a, T](t, s) \le 0 \qquad \forall\,(t, s) \in I \times I.$$

We also deduce the following result.

Corollary 19. *If $G_N[a, T] \geq 0$ on $I \times I$, then*
1. $G_N[a, T](t, s) \leq 2\,G_P[\tilde{a}, 2\,T](2\,T - t, s)$ *on $I \times I$.*
2. $0 \geq G_D[a, T](t, s) \geq -2\,G_P[\tilde{a}, 2\,T](2\,T - t, s)$ *on $I \times I$.*
3. $G_N[a, T](t, s) \leq 2\,G_N[\tilde{a}, 2\,T](2\,T - t, s)$ *on $I \times I$.*
4. $0 \geq G_{M_1}[a, T](t, s) \geq -2\,G_N[\tilde{a}, 2\,T](2\,T - t, s)$ *on $I \times I$.*

In particular, $G_P[\tilde{a}, 2\,T](2\,T - t, s) \geq 0$ and $G_N[\tilde{a}, 2\,T](2\,T - t, s) \geq 0$ on $I \times I$.

Proof. The inequalities are deduced from expressions (3.25), (3.26), (3.31), and (3.32) by taking into account that if $G_N[a, T] \geq 0$ on $I \times I$ then $G_D[a, T] \leq 0$ and $G_{M_1}[a, T] \leq 0$ on $I \times I$. $\qquad\square$

3.4.7 Constant Sign for Non-Periodic Green's Functions

Now we will use the relations obtained in the previous sections in order to generalize the results related to the periodic problem.

First, in view of Lemma 10, we are going to rewrite Proposition 3 and Lemmas 29, 30, and 32, respectively, in terms of the constant sign of the Green's function.

Lemma 38. *If $G_P[a, T] < 0$ on $I \times I$ then $\int_0^T a(s)\,ds < 0$.*

Lemma 39. *If $\int_0^T a(t)\,dt \geq 0$, $a \not\equiv 0$, and $\|a_+\|_{L^\alpha[0,T]} \leq K(2\alpha^*, T)$, then $G_P[a, T] \geq 0$ on $I \times I$.*

Lemma 40. *If $a \prec 0$, then $G_P[a, T](t, s) < 0$ for all $(t, s) \in I \times I$.*

Lemma 41. *If $a \in L^\alpha(I)$, $a \not\equiv 0$ and*

$$\int_0^T a_+(s)\,ds < \frac{4}{T}, \quad \frac{\int_0^T a_+(s)\,ds}{1 - \frac{T}{4}\int_0^T a_+(s)\,ds} \leq \int_0^T a_-(s)\,ds,$$

then $G_P[a, T] < 0$ on $I \times I$.

We can also relate the discriminant $\Delta(\lambda)$, defined in Theorem 11, with the constant sign of the Green's function.

Lemma 42. *We have the following properties:*
1. $G_P[a, T] < 0$ *on $I \times I$ if and only if $\Delta(\lambda) > 2$ for all $\lambda \leq 0$.*
2. $G_P[a, T] \geq 0$ *on $I \times I$ if and only if $\Delta(\lambda) > -2$ for all $\lambda < 0$ and $\Delta(0) < 2$.*

Lemma 43. *Suppose that $\|\hat{a}_+\|_{L^\alpha[0,T]} \leq K(2\alpha^*, T)$, with $a(t) = \hat{a}(t) + \lambda$, where $\hat{a}$ has mean zero and $\lambda = \frac{1}{T}\int_0^T a(s)\,ds$ is the mean value of a. Then*

1. $G_P[a, T] < 0$ on $I \times I$ if and only if $\int_0^T a(s)\, ds < 0$ and $\Delta(0) > 2$.
2. If $\int_0^T a(s)\, ds < 0$ and $\Delta(0) < 2$, then $G_P[a, T] \geq 0$ on $I \times I$.
3. If operator $L[a]$ under periodic boundary conditions is nonresonant and

$$0 \leq \int_0^T a(s)\, ds \leq \frac{\pi^2}{T}\left(1 - \frac{\|\hat{a}_+\|_{L^\alpha[0,T]}}{K(2\alpha^*, T)}\right),$$

then $G_P[a, T] \geq 0$ on $I \times I$.

Now, as a corollary of Lemma 24, we obtain the following bounds for the first eigenvalues of Neumann, Dirichlet, and mixed problems.

Corollary 20. *Let* $a \in L^\alpha(I)$. *Then:*
1. $\lambda_0^N[a, T] \leq -\frac{1}{T}\int_0^T a(s)\, ds$ *and the equality holds if and only if a is constant.*
2. *If* $\|a_+\|_{L^\alpha[0,T]} \leq 2^{-1/\alpha} K(2\alpha^*, 2\,T)$ *then*

$$\min\left\{\lambda_0^{M_1}[a, T], \lambda_0^{M_2}[a, T]\right\} \geq \left(\frac{\pi}{2\,T}\right)^2\left(1 - \frac{2^{1/\alpha}\,\|a_+\|_{L^\alpha[0,T]}}{K(2\alpha^*, 2\,T)}\right) \geq 0$$

and, consequently,

$$\lambda_0^D[a, T] > \left(\frac{\pi}{2\,T}\right)^2\left(1 - \frac{2^{1/\alpha}\,\|a_+\|_{L^\alpha[0,T]}}{K(2\alpha^*, 2\,T)}\right) \geq 0.$$

Proof. 1.   As a consequence of item (*i*) in Lemma 24 applied to the even extension of a we know that

$$\lambda_0^P[\tilde{a}, 2\,T] \leq -\frac{1}{2\,T}\int_0^{2\,T} \tilde{a}(s)\, ds$$

and the equality holds only when $\tilde{a}$ is constant. Assertion one in Theorem 18 warrants that $\lambda_0^N[a, T] = \lambda_0^P[\tilde{a}, 2\,T]$. Then, we deduce that

$$\lambda_0^N[a, T] \leq -\frac{1}{2\,T}\int_0^{2\,T} \tilde{a}(s)\, ds = -\frac{1}{2\,T}\left(\int_0^T a(s)\, ds + \int_T^{2\,T} a(2\,T - s)\, ds\right)$$

$$= -\frac{1}{2\,T}\left(2\int_0^T a(s)\, ds\right) = -\frac{1}{T}\int_0^T a(s)\, ds$$

and the equality holds if and only if a is constant.
2.   From assertion (*ii*) in Lemma 24 we have that if $\|\tilde{a}_+\|_{L^\alpha[0,2T]} \leq K(2\alpha^*, 2\,T)$ then

$$\lambda_0^A[\tilde{a}, 2\,T] \geq \left(\frac{\pi}{2\,T}\right)^2\left(1 - \frac{\|\tilde{a}_+\|_{L^\alpha[0,2T]}}{K(2\alpha^*, 2\,T)}\right) \geq 0.$$

We also have that

$$\|\tilde{a}_+\|_{L^\alpha[0,2T]}^\alpha = \int_0^{2T} |\tilde{a}_+(s)|^\alpha \, ds = 2 \int_0^T |a_+(s)|^\alpha \, ds = 2 \, \|a_+\|_{L^\alpha[0,T]}^\alpha,$$

from where we deduce that

$$\|\tilde{a}_+\|_{L^\alpha[0,2T]} = 2^{1/\alpha} \, \|a_+\|_{L^\alpha[0,T]}.$$

The result can be concluded from the fact that

$$\lambda_0^A[\tilde{a}, 2T] = \min\left\{\lambda_0^{M_1}[a, T], \lambda_0^{M_2}[a, T]\right\} < \lambda_0^D[a, T],$$

proved in Theorem 18.   □

On the other hand, using the implications between the constant sign of the Green's functions of the different problems, formulated in Corollary 15, we can rewrite Lemmas 38, 39, 40, and 41. For that, it is enough to consider those lemmas in terms of $\tilde{a}$ and take into account the following relations (which are deduced from the previous proof):

- Condition $a \succ 0$ ($\prec 0$) is equivalent to $\tilde{a} \succ 0$ ($\prec 0$).
- We have the following relation between the norms of a and $\tilde{a}$

$$\|\tilde{a}\|_{L^\alpha[0,2T]} = 2^{1/\alpha} \, \|a\|_{L^\alpha[0,T]}.$$

- As $\tilde{a}$ is an even function, it verifies that

$$\int_0^{2T} \tilde{a}(s) \, ds = 2 \int_0^T a(s) \, ds.$$

We are now in condition to rewrite in each case the hypothesis in the corresponding terms.

Corollary 21.   (i)   If $a \prec 0$, then $G_N[a, T] < 0$ on $I \times I$.
(ii)   If $\int_0^T a(t) \, dt \geq 0$, $a \not\equiv 0$, and $\|a_+\|_{L^\alpha[0,T]} \leq 2^{-1/\alpha} K(2\alpha^*, 2T)$, then $G_N[a, T] \geq 0$ on $I \times I$.
(iii)   If $a \in L^\alpha(I)$, $a \not\equiv 0$ and

$$\int_0^T a_+(s) \, ds < \frac{1}{T}, \qquad \frac{\int_0^T a_+(s) \, ds}{1 - T \int_0^T a_+(s) \, ds} \leq \int_0^T a_-(s) \, ds,$$

then $G_N[a, T] < 0$ on $I \times I$.

Any of the previous conditions implies that:

1. $G_{M_1}[a, T](t, s) < 0$ for all $(t, s) \in [0, T) \times [0, T)$.
2. $G_{M_2}[a, T](t, s) < 0$ for all $(t, s) \in (0, T] \times (0, T]$.
3. $G_D[a, T](t, s) < 0$ for all $(t, s) \in (0, T) \times (0, T)$.
4. $G_D[\tilde{a}, 2\,T](t, s) < 0$ for all $(t, s) \in (0, 2\,T) \times (0, 2\,T)$.

Corollary 22. *If $G_N[a, T] < 0$ on $I \times I$ then $\int_0^T a(s)\, ds < 0$.*

Considering now the discriminant $\tilde{\Delta}(\lambda) = u_1(2\,T, \lambda) + u_2'(2\,T, \lambda)$ for the periodic problem on $[0, 2\,T]$ with potential $\tilde{a}$, it is possible to obtain results about the constant sign of non-periodic Green's functions from Lemmas 42 and 43.

Corollary 23. *The following properties hold:*

(i) $G_N[a, T] < 0$ on $I \times I$ if and only if $\tilde{\Delta}(\lambda) > 2$ for all $\lambda \leq 0$.
(ii) $G_N[a, T] \geq 0$ on $I \times I$ if and only if $\tilde{\Delta}(\lambda) > -2$ for all $\lambda < 0$ and $\tilde{\Delta}(0) < 2$.

Corollary 24. *If $\tilde{\Delta}(\lambda) > -2$ for all $\lambda < 0$ then:*

1. $G_{M_1}[a, T](t, s) < 0$ for all $(t, s) \in [0, T) \times [0, T)$.
2. $G_{M_2}[a, T](t, s) < 0$ for all $(t, s) \in (0, T] \times (0, T]$.
3. $G_D[a, T](t, s) < 0$ for all $(t, s) \in (0, T) \times (0, T)$.
4. $G_D[\tilde{a}, 2\,T](t, s) < 0$ for all $(t, s) \in (0, 2\,T) \times (0, 2\,T)$.

Corollary 25. *Suppose that $\left\| \hat{a}_+ \right\|_{L^\alpha[0, T]} \leq 2^{-1/\alpha} K(2\alpha^*, 2\,T)$, with $a(t) = \hat{a}(t) + \lambda$, where $\hat{a}$ has mean zero and $\lambda = \frac{1}{T} \int_0^T a(s)\, ds$ is the mean value of a. Then:*

(i) $G_N[a, T] < 0$ on $I \times I$ if and only if $\int_0^T a(s)\, ds < 0$ and $\tilde{\Delta}(0) > 2$.
(ii) If $\int_0^T a(s)\, ds < 0$ and $\tilde{\Delta}(0) < 2$, then $G_N[a, T] \geq 0$ on $I \times I$.
(iii) If operator $L[a]$ under periodic boundary conditions on $[0, 2\,T]$ is nonresonant and

$$0 \leq \int_0^T a(s)\, ds \leq \frac{\pi^2}{4\,T} \left(1 - \frac{2^{1/\alpha} \left\| \hat{a}_+ \right\|_{L^\alpha[0, T]}}{K(2\alpha^*, 2\,T)} \right),$$

then $G_N[a, T] \geq 0$ on $I \times I$.

Corollary 26. *If either*

$$\int_0^T a(s)\, ds < 0$$

or

$$0 \leq \int_0^T a(s)\, ds \leq \frac{\pi^2}{4\,T} \left(1 - \frac{2^{1/\alpha} \left\| \hat{a}_+ \right\|_{L^\alpha[0, T]}}{K(2\alpha^*, 2\,T)} \right),$$

then:

1. $G_{M_1}[a, T](t, s) < 0$ for all $(t, s) \in [0, T) \times [0, T)$.
2. $G_{M_2}[a, T](t, s) < 0$ for all $(t, s) \in (0, T] \times (0, T]$.
3. $G_D[a, T](t, s) < 0$ for all $(t, s) \in (0, T) \times (0, T)$.
4. $G_D[\tilde{a}, 2\,T](t, s) < 0$ for all $(t, s) \in (0, 2\,T) \times (0, 2\,T)$.

Obviously all the results in this section could be formulated in terms of the maximum and antimaximum principles.

3.4.8 Global Order of Eigenvalues

It can also be proved that there exists a certain order for the eigenvalues related to (N, T), (D, T), (M_1, T), and (M_2, T).

Consider the following facts:

(i) Let $\lambda_k^N[a, T]$, $\lambda_{k+1}^N[a, T] \in \Lambda_N[a, T]$ be two consecutive eigenvalues of Neumann problem (N, T) and let $u_k^{N,T}$ and $u_{k+1}^{N,T}$ be their associated eigenfunctions, with k and $k + 1$ zeros on the interval $[0, T]$, respectively.

If we consider the even extensions of $u_k^{N,T}$ and $u_{k+1}^{N,T}$ to the interval $[0, 2\,T]$, it is clear that they have $2k$ and $2k + 2$ zeros on $[0, 2\,T]$, respectively, so there must exist an eigenvalue $\lambda \in \Lambda_N[\tilde{a}, 2\,T]$, $\lambda_k^N[a, T] < \lambda < \lambda_{k+1}^N[a, T]$, such that its associated eigenfunction has exactly $2k + 1$ zeros on the interval $[0, 2\,T]$. From the decomposition of the Neumann spectrum showed in Subsection 3.4.4, we have that, necessarily, $\lambda \in \Lambda_{M_1}[a, T]$.

As we know that $\lambda_0^N[\tilde{a}, 2\,T] = \lambda_0^N[a, T]$ we conclude that

$$\cdots < \lambda_k^N[a, T] < \lambda_k^{M_1}[a, T] < \lambda_{k+1}^N[a, T] < \lambda_{k+1}^{M_1}[a, T] < \ldots$$

(ii) Analogously, we can easily see that $\Lambda_{M_2}[a, T]$ corresponds with the eigenvalues of $\Lambda_D[\tilde{a}, 2\,T]$ whose eigenfunctions have an even number of zeros on $(0, 2\,T)$ and $\Lambda_D[a, T]$ corresponds with the eigenvalues of $\Lambda_D[\tilde{a}, 2\,T]$ whose eigenfunctions have an odd number of zeros on $(0, 2\,T)$. Taking into account the fact that $\lambda_0^D[\tilde{a}, 2\,T] = \lambda_0^{M_2}[a, T]$ we conclude that

$$\cdots < \lambda_k^{M_2}[a, T] < \lambda_k^D[a, T] < \lambda_{k+1}^{M_2}[a, T] < \lambda_{k+1}^D[a, T] < \ldots$$

(iii) Oscillation Theorem (Theorem 11) guarantees that the eigenvalues of periodic and antiperiodic problems related to the same interval always appear in the following order

$$\lambda_0^P[a, T] < \lambda_0^A[a, T] \leq \lambda_1^A[a, T] < \lambda_1^P[a, T] \leq \lambda_2^P[a, T] < \lambda_2^A[a, T]$$
$$\leq \lambda_3^A[a, T] < \ldots$$

Consequently, if we consider item (*iii*) for problems $(P, 2T)$ and $(A, 2T)$ and we take into account the inequalities obtained in items (*i*) and (*ii*) we can affirm that

- In each pair $\{\lambda_{2k-1}^P[a, 2T], \lambda_{2k}^P[a, 2T]\}$ of two consecutive eigenvalues of problem $(P, 2T)$, one of them belongs to $\Lambda_N[a, T]$ and the other one belongs to $\Lambda_D[a, T]$.
- In each pair $\{\lambda_{2k}^A[a, 2T], \lambda_{2k+1}^A[a, 2T]\}$ of two consecutive eigenvalues of problem $(A, 2T)$, one of them belongs to $\Lambda_{M_1}[a, T]$ and the other one belongs to $\Lambda_{M_2}[a, T]$.

The previous reasoning lets us conclude that the eigenvalues of problem $(P, 4T)$ always appear in the following order:

$$\lambda_0^N[a, T] < \left\{\lambda_0^{M_1}[a, T], \ \lambda_0^{M_2}[a, T]\right\} < \left\{\lambda_0^D[a, T], \ \lambda_1^N[a, T]\right\}$$
$$< \left\{\lambda_1^{M_1}[a, T], \ \lambda_1^{M_2}[a, T]\right\}$$
$$< \left\{\lambda_1^D[a, T], \ \lambda_2^N[a, T]\right\} < \ldots$$

As an immediate consequence we deduce

Corollary 27. *The following properties hold for any $a \in L^\alpha(I)$.*
1. $\lambda_k^N[a, T] < \lambda_k^{M_2}[a, T] < \lambda_{k+1}^N[a, T] < \lambda_{k+1}^{M_2}[a, T], \quad k = 0, 1, \ldots$
2. $\lambda_k^{M_1}[a, T] < \lambda_k^D[a, T] < \lambda_{k+1}^{M_1}[a, T] < \lambda_{k+1}^D[a, T], \quad k = 0, 1, \ldots$

Remark 28. In [22, Chapter 1] the following equalities are proved in the case of an even potential on $[0, 2T]$:

$$u_1(2T, \lambda) = 2u_1(T, \lambda)u_2'(T, \lambda) - 1 = 1 + 2u_1'(T, \lambda)u_2(T, \lambda), \quad (3.41)$$
$$u_1'(2T, \lambda) = 2u_1(T, \lambda)u_1'(T, \lambda), \quad (3.42)$$
$$u_2(2T, \lambda) = 2u_2(T, \lambda)u_2'(T, \lambda), \quad (3.43)$$
$$u_2'(2T, \lambda) = u_1(2T, \lambda), \quad (3.44)$$

with u_1 and u_2 the fundamental solutions of Eq. (1.9), given in Definition 4.

On the other hand we have proved in Chapter 2 that

- $\lambda \in \Lambda_N[a, T]$ if and only if $u_1'(T, \lambda) = 0$,
- $\lambda \in \Lambda_D[a, T]$ if and only if $u_2(T, \lambda) = 0$,
- $\lambda \in \Lambda_{M_1}[a, T]$ if and only if $u_1(T, \lambda) = 0$,
- $\lambda \in \Lambda_{M_2}[a, T]$ if and only if $u_2'(T, \lambda) = 0$.

Therefore we deduce that, as $\tilde{a}$ is an even function, the decomposition of Neumann and Dirichlet spectrum in $2\,T$,

$$\Lambda_N[\tilde{a}, 2\,T] = \Lambda_N[a, T] \cup \Lambda_{M_1}[a, T]$$

and

$$\Lambda_D[\tilde{a}, 2\,T] = \Lambda_D[a, T] \cup \Lambda_{M_2}[a, T],$$

could also be deduced from the equalities (3.42) and (3.43).

3.4.9 Examples

We will see now some examples of the different situations that we could find. To calculate the eigenvalues we will use the characterization of the spectrum given in Remark 28.

Example 6. If we consider the constant case $a(t) = 0$, it is known that (see [4])

$$\lambda_0^P[0, 2\,T] = \lambda_0^N[0, T] = 0 \quad \text{and} \quad \lambda_0^A[0, 2\,T] = \lambda_0^D[0, 2\,T] = \left(\frac{\pi}{2\,T}\right)^2.$$

Moreover, denoting $\lambda = m^2 > 0$ and using [8] we obtain

$$G_P[m^2, 2\,T](t, s) = \frac{1}{2\,m \sin m\,T} \begin{cases} \cos(m\,(s - t + T)), & 0 \le s \le t \le 2T \\ \cos(m\,(s - t - T)), & 0 \le t \le s \le 2T \end{cases}$$

and

$$G_N[m^2, T](t, s) = \frac{1}{m \sin m\,T} \begin{cases} \cos(m\,s) \cos(m\,(T - t)), & 0 \le s \le t \le T \\ \cos(m\,t) \cos(m\,(T - s)), & 0 \le t \le s \le T \end{cases}.$$

It is obvious that

$$G_N[m^2, T](0, 0) = 2\,G_P[m^2, 2\,T](0, 0) = \frac{1}{m \tan m\,T}.$$

As a consequence we know that $\lambda_0^{M_1}[0, T] = \left(\frac{\pi}{2\,T}\right)^2$.
From the fact that

$$G_N[m^2, T](T, T) = 2\,G_P[m^2, 2\,T](T, T) = \frac{1}{m \tan m\,T},$$

we deduce that $\lambda_0^{M_2}[0, T] = \left(\frac{\pi}{2\,T}\right)^2$. This is also deduced from Corollary 10.

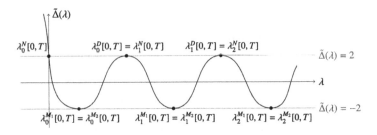

Figure 3.7 Graphic of $\tilde{\Delta}(\lambda)$ for $a(t) = 0$.

We can use [8] to calculate the Green's functions for the different boundary conditions

$$G_D[m^2, T](t, s) = \frac{1}{m \sin m T} \begin{cases} \sin(ms) \sin(m(t-T)), & 0 \le s \le t \le T \\ \sin(mt) \sin(m(s-T)), & 0 \le t \le s \le T \end{cases},$$

$$G_{M_1}[m^2, T](t, s) = \frac{1}{m \cos m T} \begin{cases} \cos(ms) \sin(m(t-T)), & 0 \le s \le t \le T \\ \cos(mt) \sin(m(s-T)), & 0 \le t \le s \le T \end{cases},$$

$$G_{M_2}[m^2, T](t, s) = \frac{-1}{m \cos m T} \begin{cases} \sin(ms) \cos(m(T-t)), & 0 \le s \le t \le T \\ \sin(mt) \cos(m(T-s)), & 0 \le t \le s \le T \end{cases}$$

and

$$G_A[m^2, 2T](t, s) = \frac{-1}{2m \cos m T} \begin{cases} \sin(m(s-t+T)), & 0 \le s \le t \le T \\ \sin(m(-s+t+T)), & 0 \le t \le s \le T \end{cases}.$$

We observe then that $\lambda_0^D[0, T] = \left(\frac{\pi}{T}\right)^2$.

In this case $\Lambda_N[a, T] = \Lambda_D[a, T] \cup \{0\} = \Lambda_P[\tilde{a}, 2T]$ and $\Lambda_{M_1}[a, T] = \Lambda_{M_2}[a, T] = \Lambda_A[\tilde{a}, 2T]$.

Then, if we represent $\tilde{\Delta}(\lambda) = u_1(2T, \lambda) + u_2'(2T, \lambda)$ graphically, we obtain Fig. 3.7.

Example 7. If we consider $T = 2$, and $a(t) = 0$, for $t \in [0, 1]$ and $a(t) = 1/10$, for $t \in [1, 2]$, the eigenvalues can be directly obtained and we can verify that

$$\lambda_0^N[a, 2] = \lambda_0^N[\tilde{a}, 4] = \lambda_0^P[\tilde{a}, 4] \approx -0.0508,$$
$$\lambda_0^{M_2}[a, 2] = \lambda_0^D[\tilde{a}, 4] = \lambda_0^A[\tilde{a}, 4] \approx 0.5346,$$
$$\lambda_0^{M_1}[a, 2] \approx 0.5984$$

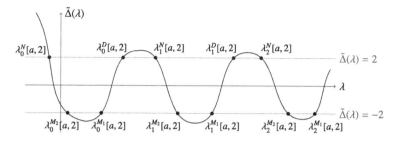

Figure 3.8 Graphic of $\tilde{\Delta}(\lambda)$ for a piecewise constant potential a.

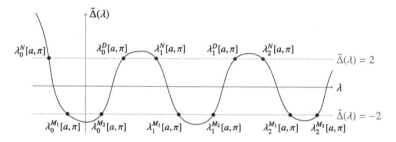

Figure 3.9 Graphic of $\tilde{\Delta}(\lambda)$ for $a(t) = \cos t$.

and

$$\lambda_0^D[a, 2] \approx 2.4170.$$

Graphically, the situation would be represented in Fig. 3.8.

Note that the eigenvalue of problem (M_2, T) always appears before the one of problem (M_1, T). In addition, the order between the eigenvalues of (N, T) and (D, T) is also maintained.

Example 8. Considering $T = \pi$ and $a(t) = \cos t$, we obtain the following approximations

$$\lambda_0^N[a, \pi] = \lambda_0^N[\tilde{a}, 2\pi] = \lambda_0^P[\tilde{a}, 2\pi] = \lambda_0^P[\tilde{\tilde{a}}, 4\pi] \approx -0.378,$$
$$\lambda_0^{M_1}[a, \pi] = \lambda_0^A[\tilde{a}, 2\pi] \approx -0.348,$$
$$\lambda_0^{M_2}[a, \pi] = \lambda_0^D[\tilde{a}, 2\pi] \approx 0.5948$$

and

$$\lambda_0^D[a, \pi] \approx 0.918.$$

Graphically we would obtain Fig. 3.9.

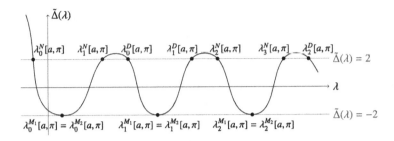

Figure 3.10 Graphic of $\tilde{\Delta}(\lambda)$ for $a(t) = \cos 2t$.

In this case, the eigenvalue of (M_1, T) is smaller than the one of (M_2, T). Again, the order between the eigenvalues of (N, T) and (D, T) is maintained.

The following example shows that eigenvalues related to problem (N, T) do not necessarily have to alternate with the ones related to (D, T).

Example 9. Considering $T = \pi$ and $a(t) = \cos 2t$, we obtain the following approximation for the spectrum of the considered problems

$$\Lambda_P[\tilde{a}, 4\pi] = \{-0.1218, 0.0923, 0.47065, 1.4668, 2.34076, 3.9792,$$
$$4.1009, \dots\},$$
$$\Lambda_P[\tilde{a}, 2\pi] = \{-0.1218, 0.47065, 1.4668, 3.9792, 4.1009, \dots\},$$
$$\Lambda_{M_1}[a, \pi] = \Lambda_{M_2}[a, \pi] = \Lambda_A[\tilde{a}, 2\pi] = \{0.0923, 2.34076, \dots\},$$
$$\Lambda_N[a, \pi] = \{-0.1218, 0.47065, 4.1009, \dots\}$$

and

$$\Lambda_D[a, \pi] = \{1.4668, 3.9792, \dots\}.$$

We observe that in this case

$$\lambda_0^N[a, \pi] < \lambda_1^N[a, \pi] < \lambda_0^D[a, \pi] < \lambda_1^D[a, \pi] < \lambda_2^N[a, \pi].$$

Note that the eigenvalues of mixed problems coincide. This is due to the fact that $a(t) = a(\pi - t)$. Consequently, all the eigenvalues of $\Lambda_A[\tilde{a}, 2\pi]$ are a double root of $\tilde{\Delta}(\lambda) = -2$.

Graphically we get Fig. 3.10.

Remark 29. The numerical results obtained in the considered examples suggest an order of eigenvalues even more precise than the one theoretically proved.

It is observed that the eigenvalues of mixed problems alternate, with one eigenvalue of a mixed problem between two consecutive eigenvalues of the other one, and reciprocally. This has been observed in all the considered examples in which the spectrum of the two mixed problems are different (Examples 7 and 8), independently of which of them appears first.

We also appreciate in the examples an alternation between Neumann and Dirichlet eigenvalues except for the case in which the spectrum of the mixed problems is the same (in this case the order of appearance of Dirichlet and Neumann changes between one pair of eigenvalues and the next one, as we can see in Example 9).

This situation suggests the existence of some property justifying this fact. However, for the moment this has not been formally proved and these speculations are uniquely based on the numerical results obtained while working with different potentials.

3.5 GENERAL SECOND ORDER EQUATION

In this section a more general problem is considered and it is proved to be equivalent to the one previously treated. As a consequence, all the results obtained until the moment could be rewritten in terms of this problem. First, the periodic problem will be studied and later the obtained results will be generalized to other boundary conditions.

3.5.1 Periodic Problem

Consider now the general second order equation given in self-adjoint form

$$\begin{cases} (p\,u')'(t) + \bar{a}(t)\,u(t) = \bar{\sigma}(t), & \text{a.e.} \quad t \in I, \\ u(0) = u(T), \\ (p\,u')(0) = (p\,u')(T), \end{cases} \tag{3.45}$$

with $p > 0$ a.e. $t \in I$ and $\frac{1}{p} \in L^1(I)$.

Let $\bar{a}$ and $\bar{\sigma}$ be such that $\bar{a}\,p^{\frac{\alpha-1}{\alpha}}$ and $\bar{\sigma}\,p^{\frac{\alpha-1}{\alpha}} \in L^\alpha(I)$, for some $\alpha \in [1, \infty]$. Let $u \in AC(I)$ be a solution of problem (3.45) such that $p\,u' \in AC(I)$.

Defining

$$w(t) = \int_0^t \frac{ds}{p(s)}, \quad t \in I,$$

we have that $w \in AC(I)$ and $w'(t) = \frac{1}{p(t)} > 0$ a.e. $t \in I$. Moreover, $w(0) = 0$ and $w(T) = R > 0$.

As a consequence, both w and w^{-1} are continuous and strictly increasing functions on their intervals of definition.

Let $y : [0, R] \equiv K \to \mathbb{R}$ be defined as $y(t) = u(w^{-1}(t))$. Obviously, $y \in C(K)$ and, since $u \in AC(I)$ and $w^{-1} : K \to I$ is a monotone function, the following theorem warrants that $y = u \circ w^{-1} \in AC(K)$.

Theorem 24. *[23, Theorem 9.3] Let $g \in AC([a, b])$ be such that $g([a, b]) \subset [\alpha, \beta]$ and $f \in AC([\alpha, \beta])$. If any of the two following conditions is satisfied*
(i) g is monotone,
(ii) f is Lipschitz,
then $f \circ g \in AC([a, b])$.

Moreover, if $f \in AC([\alpha, \beta])$ and $g \in AC([a, b])$ is monotone, it is verified that

$$(f \circ g)'(t) = f'(g(t)) g'(t) \quad \text{a.e. } t \in [a, b].$$

This result can be seen in [10, Remark 3] and is deduced from [23, Theorems 9.3 and 38.4].

Therefore, for a.e. $t \in K$, the following equality is satisfied:

$$y'(t) = u'(w^{-1}(t)) (w^{-1}(t))' = u'(w^{-1}(t)) \frac{1}{w'(w^{-1}(t))} = (p u') (w^{-1}(t)).$$

In an analogous way it can be deduced that $(p u') \circ w^{-1} \in AC(K)$ and

$$y''(t) = ((p u') \circ w^{-1})'(t) = (p u')' (w^{-1}(t)) p(w^{-1}(t)) \in L^1(I).$$

Consequently, we have that for a.e. $t \in K$

$$y''(t) + p(w^{-1}(t)) \bar{a}(w^{-1}(t)) y(t) = p(p u')' (w^{-1}(t)) + (p \bar{a}) (w^{-1}(t)) u(w^{-1}(t))$$
$$= (p \bar{\sigma}) (w^{-1}(t))$$

and moreover

$$y(0) = u(w^{-1}(0)) = u(0) = u(T) = u(w^{-1}(R)) = y(R),$$
$$y'(0) = \lim_{t \to 0^+} y'(t) = \lim_{t \to 0^+} (p u') (w^{-1}(t)) = \lim_{s \to 0^+} (p u')(s) = (p u')(0)$$

and

$$y'(R) = \lim_{t \to R^-} y'(t) = \lim_{t \to R^-} (p u') (w^{-1}(t)) = \lim_{s \to T^-} (p u')(s) = (p u')(T).$$

On the other hand, note that

$$
\begin{aligned}
\int_0^R \left| p(w^{-1}(t))\, \bar{a}(w^{-1}(t)) \right|^\alpha dt
&= \int_0^R \left| \bar{a}\, p^{\frac{\alpha-1}{\alpha}} (w^{-1}(t)) \right|^\alpha p(w^{-1}(t))\, dt \\
&= \int_0^R \left| \bar{a}\, p^{\frac{\alpha-1}{\alpha}} (w^{-1}(t)) \right|^\alpha (w^{-1}(t))'\, dt \\
&= \int_{w^{-1}(0)}^{w^{-1}(R)} \left| \bar{a}\, p^{\frac{\alpha-1}{\alpha}} (s) \right|^\alpha ds = \int_0^T \left| \bar{a}\, p^{\frac{\alpha-1}{\alpha}} (s) \right|^\alpha ds \\
&= \left\| \bar{a}\, p^{\frac{\alpha-1}{\alpha}} \right\|_{L^\alpha[0,T]}^\alpha < +\infty,
\end{aligned}
$$

that is,

$$
a(t) \equiv p(w^{-1}(t))\, \bar{a}(w^{-1}(t)) \in L^\alpha(K).
$$

Remark 30. The fact that a is measurable is deduced from both $p \circ w^{-1}$ and $\bar{a} \circ w^{-1}$ being measurable. Indeed, we will see that if V is open, then $(p \circ w^{-1})^{-1}(V) = w(p^{-1}(V))$ is a measurable set.

As p is a measurable function, then $p^{-1}(V)$ is measurable so it is enough to verify that w takes measurable sets into measurable sets.

Since w is absolutely continuous, the following theorem warrants that the image through w of any set with measure zero has measure zero.

Theorem 25 (Banach-Zarecki). *[19, Theorem 18.25] A function f is absolutely continuous on an interval $[a, b]$ if and only if the two following conditions are verified:*

(i) *f is continuous and of bounded variation on $[a, b]$.*

(ii) *The image through f of any subset of $[a, b]$ with measure zero is a set with measure zero.*

Moreover, as a consequence of [28, Chapter 6, Exercise 6] we have that, as w is a continuous function, the image of a set with measure zero has measure zero if and only if the image of any measurable set is measurable.

Consequently, $w(p^{-1}(V))$ is a measurable set and the function $p \circ w^{-1}$ is measurable.

An analogous reasoning could be considered for $\bar{a} \circ w^{-1}$.

Similarly we obtain that $\sigma(t) \equiv p(w^{-1}(t))\, \bar{\sigma}(w^{-1}(t)) \in L^\alpha(K)$.

As a consequence, $y \in W^{2,1}(K)$ is a solution of

$$
\begin{cases}
y''(t) + a(t)\, y(t) = \sigma(t), & \text{a.e. } t \in K, \\
y(0) = y(R), \\
y'(0) = y'(R),
\end{cases}
\tag{3.46}
$$

with $a, \sigma \in L^\alpha(K)$.

Reciprocally, let $a, \sigma \in L^{\alpha}(K)$ be arbitrary and let $y \in W^{2,1}(K)$ be a solution of problem (3.46). Consider functions p and w in the previous conditions.

Defining $u : I \to \mathbb{R}$ as $u(t) := y(w(t))$ and using the fact that $y \in AC(K)$ and $w : I \to K$ is monotone, we deduce from Theorem 24 that $u = y \circ w \in AC(I)$.

Therefore, applying again the chain rule, we have that

$$u'(t) = y'(w(t)) \, w'(t) = y'(w(t)) \, \frac{1}{p(t)} \quad \text{a.e. } t \in I,$$

that is,

$$(p \, u')(t) = y'(w(t)) \quad \text{a.e. } t \in I.$$

Since $y' \in AC(K)$ and $w : I \to J$ is monotone, we deduce again from Theorem 24 that $p \, u' = y' \circ w \in AC(I)$, and

$$p \, (p \, u')'(t) = y''(w(t)) \quad \text{a.e. } t \in I.$$

Moreover

$$(p \, u')'(t) + \bar{a}(t) \, u(t) = \frac{y''(w(t))}{p(t)} + \bar{a}(t) \, y(w(t)) = \bar{\sigma}(t) \quad \text{a.e. } t \in I,$$

with $\bar{a}(t) = \frac{a(w(t))}{p(t)}$ and $\bar{\sigma}(t) = \frac{\sigma(w(t))}{p(t)}$, $t \in I$.

Obviously

$$u(0) = y(w(0)) = y(0) = y(R) = y(w(T)) = u(T). \tag{3.47}$$

The monotony assumptions on function w let us affirm that

$$p \, u'(0) = \lim_{t \to 0^+} (p \, u')(t) = \lim_{t \to 0^+} y'(w(t)) = \lim_{s \to 0^+} y'(s) = y'(0) \tag{3.48}$$

and

$$p \, u'(T) = \lim_{t \to T^-} (p \, u')(t) = \lim_{t \to T^-} y'(w(t)) = \lim_{s \to R^-} y'(s) = y'(R). \tag{3.49}$$

Finally, we observe that

$$
\int_0^T \left| p^{\frac{\alpha-1}{\alpha}}(t)\, \bar{a}(t) \right|^\alpha dt = \int_0^T \left| a(w(t))\, p^{\frac{-1}{\alpha}}(t) \right|^\alpha dt = \int_0^T |a(w(t))|^\alpha \frac{dt}{p(t)}
$$

$$
= \int_0^T |a(w(t))|^\alpha\, w'(t)\, dt = \int_{w(0)}^{w(T)} |a(s)|^\alpha\, ds = \int_0^R |a(s)|^\alpha\, ds
$$

$$
= \|a\|_{L^\alpha[0,R]}^\alpha < +\infty.
$$

(3.50)

As a consequence $u \in AC(I)$, with $p\,u' \in AC(I)$, is a solution of problem (3.46) with $p^{\frac{\alpha-1}{\alpha}}\,\bar{a},\ p^{\frac{\alpha-1}{\alpha}}\,\bar{\sigma} \in L^\alpha(I)$.

We have proved that problems (3.45) and (3.46) are equivalent and the qualitative properties of the solutions of both problems are the same.

We will see next the relation between the corresponding Green's functions (see [5]).

Lemma 44. *Let* $\bar{G}_P[\bar{a}, T]$ *and* $G_P[a, R]$ *be the Green's functions related to problems (3.45) and (3.46), respectively. It is verified that*

$$
\bar{G}_P[\bar{a}, T](t, s) = G_P[a, R](w(t), w(s)) \quad \forall\, (t, s) \in I \times I.
$$

Proof. If u is the unique solution of problem (3.45), then, as we have proved before, $y(t) = u(w^{-1}(t))$ is the unique solution of (3.46) for $\sigma(t) = p(w^{-1}(t))\,\bar{\sigma}(w^{-1}(t))$ and satisfies that

$$
y(t) = \int_0^R G_P[a, R](t, s)\, p(w^{-1}(s))\, \bar{\sigma}(w^{-1}(s))\, ds.
$$

Conversely, if y is the unique solution of (3.46) then, using previous arguments again, $u = y(w(t))$ is the unique solution of problem (3.45). Consequently

$$
u(t) = y(w(t)) = \int_0^R G_P[a, R](w(t), s)\, p(w^{-1}(s))\, \bar{\sigma}(w^{-1}(s))\, ds
$$

$$
= \int_0^T G_P[a, R](w(t), w(s))\, \bar{\sigma}(s)\, ds,
$$

from where we deduce the result. ☐

As an immediate consequence we have

Corollary 28. 1. $\bar{G}_P[\bar{a}, T] \geq 0$ on $I \times I$ if and only if $G_P[a, R] \geq 0$ on $K \times K$.

2. $\bar{G}_P[\bar{a}, T] < 0$ on $I \times I$ if and only if $G_P[a, R] < 0$ on $K \times K$.

The previous result lets us obtain some criteria about the constant sign of the Green's function of problem (3.45) from Lemmas 38, 39, 40, and 41. These results are deduced by simply taking into account the following facts:

- As a consequence from (3.50) we have that

$$\|a\|_{L^\alpha[0,R]} = \left\| p^{\frac{\alpha-1}{\alpha}} \bar{a} \right\|_{L^\alpha[0,T]}.$$

- It is verified that

$$\int_0^R a(t) \, dt = \int_0^R p(w^{-1}(t)) \, \bar{a}(w^{-1}(t)) \, dt = \int_0^R \bar{a}(w^{-1}(t)) \, (w^{-1}(t))' \, dt$$
$$= \int_0^T \bar{a}(s) \, ds.$$

- Since by hypothesis $p > 0$ a.e. $t \in I$, condition $a \succ 0$ ($\prec 0$) is equivalent to $\bar{a} \succ 0$ ($\prec 0$). Moreover,

$$a_+(t) = p(w^{-1}(t)) \, \bar{a}_+(w^{-1}(t)), \qquad a_-(t) = p(w^{-1}(t)) \, \bar{a}_-(w^{-1}(t))$$

and consequently

$$\int_0^R a_+(t) \, dt = \int_0^T \bar{a}_+(t) \, dt \quad \text{and} \quad \int_0^R a_-(t) \, dt = \int_0^T \bar{a}_-(t) \, dt.$$

Corollary 29. If $\bar{a} \prec 0$, then $\bar{G}_P[\bar{a}, T] < 0$ on $I \times I$.

Corollary 30. If $\bar{a} \succ 0$ and $\left\| p^{\frac{\alpha-1}{\alpha}} \bar{a} \right\|_{L^\alpha[0,T]} \leq K(2\alpha^*, R)$, then $\bar{G}_P[\bar{a}, T] \geq 0$ on $I \times I$.

Corollary 31. If $\int_0^T \bar{a}(t) \, dt \geq 0$, $\bar{a} \not\equiv 0$, and $\left\| p^{\frac{\alpha-1}{\alpha}} \bar{a}_+ \right\|_{L^\alpha[0,T]} \leq K(2\alpha^*, R)$, then $\bar{G}_P[\bar{a}, T] \geq 0$ on $I \times I$.

Corollary 32. If $\bar{a} \in L^1(I)$, $\bar{a} \not\equiv 0$ and

$$\int_0^T \bar{a}_+(s) \, ds < \frac{4}{R}, \qquad \frac{\int_0^T \bar{a}_+(s) \, ds}{1 - \frac{R}{4} \int_0^T \bar{a}_+(s) \, ds} \leq \int_0^T \bar{a}_-(s) \, ds,$$

then $\bar{G}_P[\bar{a}, T] < 0$ on $I \times I$.

Corollary 33. *If* $\bar{G}_P[\bar{a}, T](t, s) < 0$ *for all* $(t, s) \in I \times I$ *then* $\int_0^T \bar{a}(s)\, ds < 0$.

Considering now the discriminant $\Delta(\lambda) = u_1(R, \lambda) + u_2'(R, \lambda)$ for problem (3.46), Lemmas 42 and 43 can be rewritten in order to obtain some conditions that assure the constant sign of the Green's function of problem (3.45).

Corollary 34. *The following properties hold*
(i) $\bar{G}_P[\bar{a}, T] < 0$ *on* $I \times I$ *if and only if* $\Delta(\lambda) > 2$ *for all* $\lambda \leq 0$.
(ii) $\bar{G}_P[\bar{a}, T] \geq 0$ *on* $I \times I$ *if and only if* $\Delta(\lambda) > -2$ *for all* $\lambda < 0$ *and* $\Delta(0) < 2$.

Corollary 35. *Suppose that* $\left\| p^{\frac{\alpha-1}{\alpha}} \hat{a}_+ \right\|_{L^\alpha[0,T]} \leq K(2\alpha^*, R)$, *with* $\bar{a}(t) = \hat{a}(t) +$
$\lambda \frac{1}{p(t)}$, *where* $\lambda = \frac{1}{R} \int_0^T \bar{a}(s)\, ds$ *is* $\frac{T}{R}$ *times the mean value of* $\bar{a}$. *Then*
(i) $\bar{G}_P[\bar{a}, T] < 0$ *on* $I \times I$ *if and only if* $\int_0^T \bar{a}(s)\, ds < 0$ *and* $\Delta(0) > 2$.
(ii)   *If* $\int_0^T \bar{a}(s)\, ds < 0$ *and* $\Delta(0) < 2$, *then* $\bar{G}_P[\bar{a}, T] \geq 0$ *on* $I \times I$.
(iii)   *If operator* $L[a]$ *under periodic boundary conditions is nonresonant and*

$$0 \leq \int_0^T \bar{a}(s)\, ds \leq \frac{\pi^2}{R} \left(1 - \frac{\left\| p^{\frac{\alpha-1}{\alpha}} \hat{a}_+ \right\|_{L^\alpha[0,T]}}{K(2\alpha^*, R)} \right),$$

then $\bar{G}_P[\bar{a}, T] \geq 0$ *on* $I \times I$.

Proof. The hypothesis of Lemma 43 (applied to $a(t) = p(w^{-1}(t))\, \bar{a}(w^{-1}(t))$) will be rewritten in terms of problem (3.45). Indeed, such result considers that $\|\hat{a}_+\|_{L^\alpha[0,R]} \leq K(2\alpha^*, R)$, with $a(t) = \hat{a}(t) + \lambda$, where $\lambda = \frac{1}{R} \int_0^R a(s)\, ds$ is the mean value of a. It is immediate to verify that

$$\lambda = \frac{1}{R} \int_0^T \bar{a}(s)\, ds$$

and, clearly, this is $\frac{T}{R}$ times the mean value of $\bar{a}$.
Moreover,

$$\|\hat{a}_+\|^\alpha_{L^\alpha[0,R]} = \|(a - \lambda)_+\|^\alpha_{L^\alpha[0,R]} = \int_0^R (a(t) - \lambda)_+^\alpha\, dt$$

$$= \int_0^R \left(p(w^{-1}(t))\, \bar{a}(w^{-1}(t)) - \lambda \right)_+^\alpha\, dt$$

$$= \int_0^T (p(s)\, \bar{a}(s) - \lambda)_+^\alpha\, \frac{1}{p(s)}\, ds = \int_0^T \left(p^{\frac{\alpha-1}{\alpha}} \left(\bar{a}(s) - \frac{\lambda}{p(s)} \right)_+ \right)^\alpha\, ds$$

$$= \left\| p^{\frac{\alpha-1}{\alpha}} \left(\bar{a} - \frac{\lambda}{p} \right)_+ \right\|^\alpha_{L^\alpha[0,T]} = \left\| p^{\frac{\alpha-1}{\alpha}} \hat{a}_+ \right\|^\alpha_{L^\alpha[0,T]},$$

that is,

$$\left\| \hat{a}_+ \right\|_{L^\alpha[0,R]} = \left\| p^{\frac{\alpha-1}{\alpha}} \hat{a}_+ \right\|_{L^\alpha[0,T]}.$$

The other changes in this corollary with respect to Lemma 43 are immediately obtained from the same considerations as in previous results. □

3.5.2 Non-Periodic Conditions

All the previous reasoning has been done considering periodic boundary conditions. Nevertheless, equalities (3.47), (3.48), and (3.49) warrant that

$$u(0) = \gamma(0), \qquad u(T) = \gamma(R),$$

and

$$(p\,u')(0) = \gamma'(0) \qquad (p\,u')(T) = \gamma'(R).$$

Consequently, periodic conditions in problems (3.45) and (3.46) can be substituted by any other kind of boundary conditions and this does not affect to the equivalence of the problems. We obtain the same relation between the Green's functions corresponding to each case, that is, using an analogous notation to the periodic case, we have the following equalities:

$$\bar{G}_N[\bar{a}, T](t, s) = G_N[a, R](w(t), w(s)) \quad \forall\, (t, s) \in I \times I,$$
$$\bar{G}_D[\bar{a}, T](t, s) = G_D[a, R](w(t), w(s)) \quad \forall\, (t, s) \in I \times I,$$
$$\bar{G}_{M_1}[\bar{a}, T](t, s) = G_{M_1}[a, R](w(t), w(s)) \quad \forall\, (t, s) \in I \times I,$$
$$\bar{G}_{M_2}[\bar{a}, T](t, s) = G_{M_2}[a, R](w(t), w(s)) \quad \forall\, (t, s) \in I \times I$$

and

$$\bar{G}_A[\bar{a}, T](t, s) = G_A[a, R](w(t), w(s)) \quad \forall\, (t, s) \in I \times I.$$

As a consequence all the results relating different Green's functions which were obtained in Subsection 3.4, are still valid in this more general case. In particular, all the corollaries in the previous section can be rewritten in terms of the Green's function of other boundary conditions, in an analogous way to Section 3.4.7. To do that it is enough to consider Lemmas 38, 39, 40, and 41 for $\tilde{a}$ (the even extension of a) and take into account the following considerations:

- We have that

$$\|\tilde{a}_+\|_{L^\alpha[0,2R]} = 2^{1/\alpha} \|a_+\|_{L^\alpha[0,R]} = 2^{1/\alpha} \left\| p^{\frac{\alpha-1}{\alpha}} \tilde{a}_+ \right\|_{L^\alpha[0,T]}.$$

- Condition $\tilde{a} \succ 0$ $(\prec 0)$ is equivalent to $a \succ 0$ $(\prec 0)$ which, at the same time, is equivalent to $\bar{a} \succ 0$ $(\prec 0)$.
- The integrals of potentials present the following relation

$$\int_0^{2R} \tilde{a}(t)\, dt = 2 \int_0^R a(t)\, dt = 2 \int_0^T \bar{a}(t)\, dt.$$

Analogously, since $p > 0$ a.e. $t \in I$,

$$\int_0^{2R} \tilde{a}_+(t)\, dt = 2 \int_0^T \bar{a}_+(t)\, dt \quad \text{and} \quad \int_0^{2R} \tilde{a}_-(t)\, dt = 2 \int_0^T \bar{a}_-(t)\, dt.$$

Corollary 36. (i) If $\bar{a} \prec 0$, then $\bar{G}_N[\bar{a}, T] < 0$ on $I \times I$.

(ii) If $\int_0^T \bar{a}(t)\, dt \geq 0$, $\bar{a} \not\equiv 0$ and $\left\| p^{\frac{\alpha-1}{\alpha}} \bar{a}_+ \right\|_{L^\alpha[0,T]} \leq 2^{-1/\alpha} K(2\alpha^*, 2R)$, then
$\bar{G}_N[\bar{a}, T] \geq 0$ on $I \times I$.

(iii) If $\bar{a} \in L^1(I)$, $\bar{a} \not\equiv 0$ and

$$\int_0^T \bar{a}_+(s)\, ds < \frac{1}{R}, \qquad \frac{\int_0^T \bar{a}_+(s)\, ds}{1 - R \int_0^T \bar{a}_+(s)\, ds} \leq \int_0^T \bar{a}_-(s)\, ds,$$

then $\bar{G}_N[\bar{a}, T] < 0$ on $I \times I$.

Any of the previous conditions implies that:
1. $\bar{G}_{M_1}[\bar{a}, T](t, s) < 0$ for all $(t, s) \in [0, T) \times [0, T)$.
2. $\bar{G}_{M_2}[\bar{a}, T](t, s) < 0$ for all $(t, s) \in (0, T] \times (0, T]$.
3. $\bar{G}_D[\bar{a}, T](t, s) < 0$ for all $(t, s) \in (0, T) \times (0, T)$.
4. $\bar{G}_D[\tilde{a}, 2T](t, s) < 0$ for all $(t, s) \in (0, 2T) \times (0, 2T)$.

Corollary 37. If $\bar{G}_N[\bar{a}, T](t, s) < 0$ for all $(t, s) \in I \times I$ then $\int_0^T \bar{a}(s)\, ds < 0$.

Moreover, considering the discriminant $\tilde{\Delta}(\lambda) = u_1(2R, \lambda) + u_2'(2R, \lambda)$ for the periodic problem

$$\begin{cases} y''(t) + \tilde{a}(t)\, y(t) = \tilde{\sigma}(t), & \text{a.e. } t \in [0, 2R], \\ y(0) = y(2R), \\ y'(0) = y'(2R), \end{cases}$$

obtained from (3.46) by simply considering $\tilde{a}$ and $\tilde{\sigma}$ the even extensions of a and σ, we can deduce results for Green's function different from the periodic one by rewriting Lemmas 42 and 43.

Corollary 38. *The following properties hold:*
(i) $\bar{G}_N[\bar{a}, T] < 0$ *on* $I \times I$ *if and only if* $\tilde{\Delta}(\lambda) > 2$ *for all* $\lambda \leq 0$.
(ii) $\bar{G}_N[\bar{a}, T] \geq 0$ *on* $I \times I$ *if and only if* $\tilde{\Delta}(\lambda) > -2$ *for all* $\lambda < 0$ *and* $\tilde{\Delta}(0) < 2$.

Corollary 39. *If* $\tilde{\Delta}(\lambda) > -2$ *for all* $\lambda < 0$ *then:*
1. $\bar{G}_{M_1}[\bar{a}, T](t, s) < 0$ *for all* $(t, s) \in [0, T) \times [0, T)$.
2. $\bar{G}_{M_2}[\bar{a}, T](t, s) < 0$ *for all* $(t, s) \in (0, T] \times (0, T]$.
3. $\bar{G}_D[\bar{a}, T](t, s) < 0$ *for all* $(t, s) \in (0, T) \times (0, T)$.
4. $\bar{G}_D[\tilde{a}, 2\, T](t, s) < 0$ *for all* $(t, s) \in (0, 2\, T) \times (0, 2\, T)$.

Corollary 40. *Suppose that* $\left\| p^{\frac{\alpha-1}{\alpha}} \hat{a}_+ \right\|_{L^\alpha[0,T]} \leq 2^{-1/\alpha} K(2\alpha^*, 2R)$, *with*
$\bar{a}(t) = \hat{\bar{a}}(t) + \lambda \frac{1}{p(t)}$, *where* $\lambda = \frac{1}{R} \int_0^T \bar{a}(s)\, ds$ *is* $\frac{T}{R}$ *times the mean value of* $\bar{a}$. *Then*
(i) $\bar{G}_N[\bar{a}, T] < 0$ *on* $I \times I$ *if and only if* $\int_0^T \bar{a}(s)\, ds < 0$ *and* $\tilde{\Delta}(0) > 2$.
(ii) *If* $\int_0^T \bar{a}(s)\, ds < 0$ *and* $\tilde{\Delta}(0) < 2$, *then* $\bar{G}_N[\bar{a}, T] \geq 0$ *on* $I \times I$.
(iii) *If operator* $L[a]$ *under periodic boundary conditions on* $[0, 2\, T]$ *is nonresonant and*

$$0 \leq \int_0^T \bar{a}(s)\, ds \leq \frac{\pi^2}{4R} \left(1 - \frac{2^{1/\alpha} \left\| p^{\frac{\alpha-1}{\alpha}} \hat{a}_+ \right\|_{L^\alpha[0,T]}}{K(2\alpha^*, 2R)} \right),$$

then $\bar{G}_N[\bar{a}, T] \geq 0$ *on* $I \times I$.

Proof. This result is obtained by applying Lemma 43 to $\tilde{a}$, the even extension of $a(t) = p(w^{-1}(t))\, \bar{a}(w^{-1}(t))$. The hypothesis of that lemma considers that $\left\| \hat{\tilde{a}}_+ \right\|_{L^\alpha[0,2R]} \leq K(2\alpha^*, 2R)$, with $\tilde{a}(t) = \hat{\tilde{a}}(t) + \lambda$, where $\lambda = \frac{1}{2R} \int_0^{2R} \tilde{a}(s)\, ds$ is the mean value of $\tilde{a}$. From the fact that $\tilde{a}$ is an even function and from the relation between a and $\tilde{a}$ we have that

$$\lambda = \frac{1}{R} \int_0^R a(s)\, ds = \frac{1}{R} \int_0^T \bar{a}(s)\, ds,$$

so λ is $\frac{T}{R}$ times the mean value of $\bar{a}$.

In addition, using again that $\tilde{a}$ is even (and so $\hat{\tilde{a}}$ is even too) and taking into account the reasoning developed in the proof of Corollary 35, we arrive at the following relation between the norms of $\hat{\tilde{a}}$ and $\hat{a}$

$$\left\| \hat{\tilde{a}}_+ \right\|_{L^\alpha[0,2R]} = 2^{1/\alpha} \left\| \hat{a}_+ \right\|_{L^\alpha[0,R]} = 2^{1/\alpha} \left\| p^{\frac{\alpha-1}{\alpha}} \hat{a}_+ \right\|_{L^\alpha[0,T]}.$$

The rest of variations in this corollary with respect to Lemma 43 are immediately deduced from the considerations used in previous results. □

Corollary 41. *If either*

$$\int_0^T \bar{a}(s)\, ds < 0$$

or

$$0 \le \int_0^T \bar{a}(s)\, ds \le \frac{\pi^2}{4R} \left(1 - \frac{2^{1/\alpha} \left\| p^{\frac{\alpha-1}{\alpha}} \hat{a}_+ \right\|_{L^\alpha[0,T]}}{K(2\alpha^*, 2R)} \right),$$

then:
1. $\bar{G}_{M_1}[\bar{a}, T](t, s) < 0$ *for all* $(t, s) \in [0, T) \times [0, T)$.
2. $\bar{G}_{M_2}[\bar{a}, T](t, s) < 0$ *for all* $(t, s) \in (0, T] \times (0, T]$.
3. $\bar{G}_D[\bar{a}, T](t, s) < 0$ *for all* $(t, s) \in (0, T) \times (0, T)$.
4. $\bar{G}_D[\tilde{a}, 2\,T](t, s) < 0$ *for all* $(t, s) \in (0, 2\,T) \times (0, 2\,T)$.

Example 10. Consider the equation

$$\left(\frac{1}{t} u'(t) \right)' + \lambda\, t\, u(t) = 0, \quad t \in [0, 1], \tag{3.51}$$

$$u(0) = u(1), \quad \lim_{t \to 0^+} (t\, u(t))'(t) = \lim_{t \to 1^-} (t\, u(t))'(t), \tag{3.52}$$

which is a periodic problem of type (3.45) with $\bar{a}(t) = \lambda\, t$, $p(t) = \frac{1}{t}$, and $[0, T] = [0, 1]$.

With the definitions given at the beginning of this section, we have that

$$w(t) = \int_0^t s\, ds = \frac{t^2}{2}, \quad t \in [0, 1]$$

and

$$w^{-1}(t) = \sqrt{2t}, \quad t \in \left[0, \frac{1}{2} \right] \equiv [0, R],$$

so the previous problem is equivalent to the constant coefficients periodic problem

$$\begin{cases} y''(t) + \lambda\, y(t) = \sigma(t), & \text{a.e. } t \in \left[0, \frac{1}{2} \right], \\ y(0) = y\left(\frac{1}{2} \right), \\ y'(0) = y'\left(\frac{1}{2} \right). \end{cases} \tag{3.53}$$

Using [8] we can calculate the Green's function related to problem (3.53). We will distinguish between the case $\lambda = m^2 > 0$ and $\lambda =$

$-m^2 < 0$ ($\lambda = 0$ is not considered since it is an eigenvalue for this problem). We obtain

$$G_P[m^2, R](t, s) = \frac{1}{2\,m\,\sin\left(\frac{m}{4}\right)} \begin{cases} \cos\left(m\left(s - t + \frac{1}{4}\right)\right), & 0 \leq s \leq t \leq \frac{1}{2} \\ \cos\left(m\left(s - t - \frac{1}{4}\right)\right), & 0 \leq t < s \leq \frac{1}{2} \end{cases}$$

and

$$G_P[-m^2, R](t, s) = \frac{1}{2\,m\,\sinh\left(\frac{m}{4}\right)} \begin{cases} \cosh\left(m\left(s - t + \frac{1}{4}\right)\right), & 0 \leq s \leq t \leq \frac{1}{2} \\ \cosh\left(m\left(s - t - \frac{1}{4}\right)\right), & 0 \leq t < s \leq \frac{1}{2} \end{cases}.$$

Lemma 44 allows us to calculate the exact expression of the Green's function related to the periodic problem (3.51), (3.52). Such function is

$$\bar{G}_P[\bar{a}, T](t, s) = \frac{1}{2\,m\,\sin\left(\frac{m}{4}\right)} \begin{cases} \cos\left(\frac{m}{2}\left(s^2 - t^2 + \frac{1}{2}\right)\right), & 0 \leq s \leq t \leq 1 \\ \cos\left(\frac{m}{2}\left(s^2 - t^2 - \frac{1}{2}\right)\right), & 0 \leq t < s \leq 1 \end{cases},$$

for $\lambda = m^2 > 0$ and

$$\bar{G}_P[\bar{a}, T](t, s) = \frac{1}{2\,m\,\sinh\left(\frac{m}{4}\right)} \begin{cases} \cosh\left(\frac{m}{2}\left(s^2 - t^2 + \frac{1}{2}\right)\right), & 0 \leq s \leq t \leq 1 \\ \cosh\left(\frac{m}{2}\left(s^2 - t^2 - \frac{1}{2}\right)\right), & 0 \leq t < s \leq 1 \end{cases},$$

for $\lambda = -m^2 < 0$.

Taking into account the eigenvalues calculated in Example 6 for a constant potential a, we can analyse the values of λ for which $G_P[\lambda, R]$ has constant sign. Lemma 44 warrants that

- $\bar{G}_P[\bar{a}, T](t, s) < 0$ for all $(t, s) \in [0, T] \times [0, T]$ if and only if $\lambda < 0$.
- $\bar{G}_P[\bar{a}, T](t, s) \geq 0$ for all $(t, s) \in [0, T] \times [0, T]$ if and only if $0 < \lambda \leq (2\pi)^2$.

Analogously, the expressions obtained in Example 6 for $\lambda = m^2$ (and also the ones corresponding to $\lambda = 0$ or $\lambda = -m^2$ calculated with [8]) allow us to deduce the exact expression of the Green's functions related to Eq. (3.51) under different boundary conditions.

For Neumann problem

$$\left(\frac{1}{t}u'(t)\right)' + \lambda\,t\,u(t) = 0, \quad t \in [0, 1], \quad \lim_{t \to 0^+}(t\,u(t))'(t) = \lim_{t \to 1^-}(t\,u(t))'(t) = 0,$$

we have that

$$\bar{G}_N[m^2 t, T](t, s) = \frac{1}{m \sin\left(\frac{m}{2}\right)} \begin{cases} \cos\left(m\frac{s^2}{2}\right)\cos\left(\frac{m}{2}(1 - t^2)\right), & 0 \le s \le t \le 1 \\ \cos\left(m\frac{t^2}{2}\right)\cos\left(\frac{m}{2}(1 - s^2)\right), & 0 \le t \le s \le 1 \end{cases}$$

and

$$\bar{G}_N[-m^2 t, T](t, s) = \frac{1}{m \sinh\left(\frac{m}{2}\right)} \begin{cases} \cosh\left(m\frac{s^2}{2}\right)\cosh\left(\frac{m}{2}(1 - t^2)\right), \\ \qquad 0 \le s \le t \le 1 \\ \cosh\left(m\frac{t^2}{2}\right)\cosh\left(\frac{m}{2}(1 - s^2)\right), \\ \qquad 0 \le t \le s \le 1 \end{cases},$$

and we deduce that
- $\bar{G}_N[\bar{a}, T](t, s) < 0$ for all $(t, s) \in [0, T] \times [0, T]$ if and only if $\lambda < 0$.
- $\bar{G}_N[\bar{a}, T](t, s) \ge 0$ for all $(t, s) \in [0, T] \times [0, T]$ if and only if $0 < \lambda \le \pi^2$.
  The Green's function related to Dirichlet problem

$$\left(\frac{1}{t}u'(t)\right)' + \lambda t u(t) = 0, \quad t \in [0, 1], \quad u(0) = u(1) = 0,$$

is

$$\bar{G}_D[m^2 t, T](t, s) = \frac{1}{m \sin\left(\frac{m}{2}\right)} \begin{cases} \sin\left(m\frac{s^2}{2}\right)\sin\left(\frac{m}{2}(t^2 - 1)\right), & 0 \le s \le t \le 1 \\ \sin\left(m\frac{t^2}{2}\right)\sin\left(\frac{m}{2}(s^2 - 1)\right), & 0 \le t \le s \le 1 \end{cases},$$

$$\bar{G}_D[0, T](t, s) = \frac{1}{2} \begin{cases} s^2(t^2 - 1), & 0 \le s \le t \le 1 \\ t^2(s^2 - 1), & 0 \le t \le s \le 1 \end{cases},$$

$$\bar{G}_D[-m^2 t, T](t, s) = \frac{1}{m \sinh\left(\frac{m}{2}\right)} \begin{cases} \sinh\left(m\frac{s^2}{2}\right)\sinh\left(\frac{m}{2}(t^2 - 1)\right), \\ \qquad 0 \le s \le t \le 1 \\ \sinh\left(m\frac{t^2}{2}\right)\sinh\left(\frac{m}{2}(s^2 - 1)\right), \\ \qquad 0 \le t \le s \le 1 \end{cases}$$

and we have that
- $\bar{G}_D[\bar{a}, T](t, s) < 0$ for all $(t, s) \in (0, T) \times (0, T)$ if and only if $\lambda < (2\pi)^2$.
  With regard to mixed problems M_1,

$$\left(\frac{1}{t}u'(t)\right)' + \lambda t u(t) = 0, \quad t \in [0, 1], \quad \lim_{t \to 0^+}(t u(t))'(t) = u(1) = 0,$$

and M_2,

$$\left(\frac{1}{t}u'(t)\right)' + \lambda\, t\, u(t) = 0, \quad t \in [0,1], \quad u(0) = \lim_{t \to 1^-}(t\, u(t))'(t) = 0,$$

the corresponding Green's functions are

$$\bar{G}_{M_1}[m^2\, t,\, T](t,s) = \frac{1}{m\cos\left(\frac{m}{2}\right)}\begin{cases} \cos\left(m\frac{s^2}{2}\right)\sin\left(\frac{m}{2}(t^2-1)\right), & 0 \le s \le t \le 1 \\ \cos\left(m\frac{t^2}{2}\right)\sin\left(\frac{m}{2}(s^2-1)\right), & 0 \le t \le s \le 1 \end{cases},$$

$$\bar{G}_{M_1}[0,\, T](t,s) = \frac{1}{2}\begin{cases} t^2-1, & 0 \le s \le t \le 1 \\ s^2-1, & 0 \le t \le s \le 1 \end{cases},$$

$$\bar{G}_{M_1}[-m^2\, t,\, T](t,s) = \frac{1}{m\cosh\left(\frac{m}{2}\right)}\begin{cases} \cosh\left(m\frac{s^2}{2}\right)\sinh\left(\frac{m}{2}(t^2-1)\right), \\ \qquad 0 \le s \le t \le 1 \\ \cosh\left(m\frac{t^2}{2}\right)\sinh\left(\frac{m}{2}(s^2-1)\right), \\ \qquad 0 \le t \le s \le 1 \end{cases}$$

and

$$\bar{G}_{M_2}[m^2\, t,\, T](t,s) = \frac{-1}{m\cos\left(\frac{m}{2}\right)}\begin{cases} \sin\left(m\frac{s^2}{2}\right)\cos\left(\frac{m}{2}(1-t^2)\right), & 0 \le s \le t \le 1 \\ \sin\left(m\frac{t^2}{2}\right)\cos\left(\frac{m}{2}(1-s^2)\right), & 0 \le t \le s \le 1 \end{cases},$$

$$\bar{G}_{M_2}[0,\, T](t,s) = -\frac{1}{2}\begin{cases} s^2, & 0 \le s \le t \le 1 \\ t^2, & 0 \le t \le s \le 1 \end{cases},$$

$$\bar{G}_{M_2}[-m^2\, t,\, T](t,s) = \frac{-1}{m\cosh\left(\frac{m}{2}\right)}\begin{cases} \sinh\left(m\frac{s^2}{2}\right)\cosh\left(\frac{m}{2}(1-t^2)\right), \\ \qquad 0 \le s \le t \le 1 \\ \sinh\left(m\frac{t^2}{2}\right)\cosh\left(\frac{m}{2}(1-s^2)\right), \\ \qquad 0 \le t \le s \le 1 \end{cases}.$$

In this case we have that

- $\bar{G}_{M_1}[\bar{a},\, T](t,s) < 0$ for all $(t,s) \in [0,T) \times [0,T)$ if and only if $\lambda < \pi^2$.
- $\bar{G}_{M_2}[\bar{a},\, T](t,s) < 0$ for all $(t,s) \in (0,T] \times (0,T]$ if and only if $\lambda < \pi^2$.

In addition, the expression of each Green's function lets us calculate the spectrum of the problem under the corresponding boundary conditions

$$\Lambda_P[\bar{a},\, T] = \left\{(4k\pi)^2,\, k = 0, 1, \dots\right\},$$
$$\Lambda_N[\bar{a},\, T] = \left\{(2k\pi)^2,\, k = 0, 1, \dots\right\},$$
$$\Lambda_D[\bar{a},\, T] = \left\{(2k\pi)^2,\, k = 1, 2, \dots\right\}$$

and

$$\Lambda_{M_1}[\bar{a}, T] = \Lambda_{M_2}[\bar{a}, T] = \left\{ (2k+1)^2 \pi^2, \ k = 0, 1, \dots \right\}.$$

REFERENCES

[1] I.V. Barteneva, A. Cabada, A.O. Ignatyev, Maximum and anti-maximum principles for the general operator of second order with variable coefficients, Appl. Math. Comput. 134 (2003) 173–184.

[2] V. Bevc, J.L. Palmer, C. Süsskind, On the design of the transition region of axisymmetric, magnetically focused beam valves, J. British Inst. Radio Eng. 18 (1958) 696–708.

[3] A. Cabada, The method of lower and upper solutions for second, third, fourth, and higher order boundary value problems, J. Math. Anal. Appl. 185 (1994) 302–320.

[4] A. Cabada, Green's Functions in the Theory of Ordinary Differential Equations, Springer Briefs in Math., 2014.

[5] A. Cabada, J.A. Cid, On the sign of the Green's function associated to Hill's equation with an indefinite potential, Appl. Math. Comput. 205 (2008) 303–308.

[6] A. Cabada, J.A. Cid, On comparison principles for the periodic Hill's equation, J. Lond. Math. Soc. (2) 86 (1) (2012) 272–290.

[7] A. Cabada, J.A. Cid, L. López-Somoza, Green's functions and spectral theory for the Hill's equation, Appl. Math. Comput. 286 (2016) 88–105.

[8] A. Cabada, J.A. Cid, B. Máquez-Villamarín, Computation of Green's functions for boundary value problems with Mathematica, Appl. Math. Comput. 219 (4) (2012) 1919–1936.

[9] A. Cabada, J.A. Cid, M. Tvrdý, A generalized anti-maximum principle for the periodic one-dimensional p-Laplacian with sign-changing potential, Nonlinear Anal. 72 (2010) 3436–3446.

[10] A. Cabada, R.L. Pouso, On first order discontinuous scalar differential equations, Nonlinear Stud. 6 (2) (1999) 161–170.

[11] A. Cabada, L. Saavedra, The eigenvalue characterization for the constant sign Green's functions of $(k, n - k)$ problems, Bound. Value Probl. (2016) 44, 35pp.

[12] A. Cabada, L. Saavedra, Disconjugacy characterization by means of spectral $(k, n - k)$ problems, Appl. Math. Lett. 52 (2016) 21–29.

[13] W.A. Coppel, Disconjugacy, Lecture Notes in Math., vol. 220, Springer-Verlag, Berlin, New York, 1971.

[14] C. De Coster, P. Habets, Two-Point Boundary Value Problems: Lower and Upper Solutions, Math. Sci. Eng., vol. 205, Elsevier B. V., Amsterdam, 2006.

[15] T.R. Ding, A boundary value problem for the periodic Brillouin focusing system, Acta Sci. Natru. Univ. Pekinensis 11 (1965) 31–38.

[16] T.R. Ding, Applications of Qualitative Methods of Ordinary Differential Equations, Higher Education Press, Beijing, 2004.

[17] R. Hakl, P.J. Torres, Maximum and antimaximum principles for a second order differential operator with variable coefficients of indefinite sign, Appl. Math. Comput. 217 (2011) 7599–7611.

[18] S. Heikkilä, V. Lakshmikantham, Monotone Iterative Techniques for Discontinuous Nonlinear Differential Equations, Monogr. Textb. Pure Appl. Math., vol. 181, Marcel Dekker, Inc., New York, 1994.

[19] E. Hewitt, K.R. Stromberg, Real and Abstract Analysis, Springer-Verlag, New York, 1975.

[20] W.G. Kelley, A.C. Peterson, Difference Equations. An Introduction With Applications, second edition, Harcourt/Academic Press, San Diego, CA, 2001.

[21] G.S. Ladde, V. Lakshmikantham, A.S. Vatsala, Monotone Iterative Techniques for Nonlinear Differential Equations, Pitman, Boston M.A., 1985.

[22] W. Magnus, S. Winkler, Hill's Equation, Dover Publications, New York, 1979.

[23] E.J. McShane, Integration, Princeton University Press, 1947.

[24] M.N. Nkashama, Generalized upper and lower solutions method and multiplicity results for nonlinear first-order ordinary differential equations, J. Math. Anal. Appl. 140 (1989) 381–395.

[25] P.J. Olver, Introduction to Partial Differential Equations, Undergrad. Texts Math., Springer, 2014.

[26] P. Omari, M. Trombetta, Remarks on the lower and upper solutions method for second and third-order periodic boundary value problems, Appl. Math. Comput. 50 (1992) 1–21.

[27] J. Ren, Z. Cheng, S. Siegmund, Positive periodic solution for Brillouin electron beam focusing system, Discrete Contin. Dyn. Syst. Ser. B 16 (2011) 385–392.

[28] K.R. Stromberg, An Introduction to Classical Real Analysis, Chapman & Hall, London, 1996.

[29] P.J. Torres, Existence and uniqueness of elliptic periodic solutions of the Brillouin electron beam focusing system, Math. Meth. Appl. Sci. 23 (2000) 1139–1143.

[30] P.J. Torres, Existence of one-signed periodic solutions of some second-order differential equations via a Krasnoselskii fixed point theorem, J. Differential Equations 190 (2003) 643–662.

[31] P.J. Torres, M. Zhang, A monotone iterative scheme for a nonlinear second order equation based on a generalized anti-maximum principle, Math. Nachr. 251 (2003) 101–107.

[32] H.F. Weinberger, A First Course in Partial Differential Equations With Complex Variables and Transform Methods, Dover Publications, New York, 1995.

[33] Y. Ye, X. Wang, Nonlinear differential equations in electron beam focusing theory, Acta Math. Appl. Sin. 1 (1978) 13–41.

[34] M. Zhang, Periodic solutions of Liénard equations with singular forces of repulsive type, J. Math. Anal. Appl. 203 (1996) 254–269.

[35] M. Zhang, A relationship between the periodic and the Dirichlet BVPs of singular differential equations, Proc. Roy. Soc. Edinbourgh Sect. A 128 (1998) 1099–1114.

[36] M. Zhang, Optimal conditions for maximum and anti-maximum principles of the periodic solution problem, Bound. Value Probl. (2010) 410986, http://dx.doi.org/10.1155/2010/410986, 26pp.

[37] M. Zhang, W. Li, A Lyapunov-type stability criterion using L^{α} norms, Proc. Amer. Math. Soc. 130 (2002) 3325–3333.

CHAPTER 4

Nonlinear Equations

Contents

4.1	Introduction	129
4.2	Fixed Point Theorems and Degree Theory	130
	4.2.1 Leray-Schauder Degree	131
	4.2.2 Fixed Point Theorems	134
	4.2.3 Extremal Fixed Points	148
	4.2.4 Monotone Operators	150
	4.2.5 Non-increasing Operators	159
	4.2.6 Non-decreasing Operators	163
	4.2.7 Problems with Parametric Dependence	168
4.3	Lower and Upper Solutions Method	184
	4.3.1 Well Ordered Lower and Upper Solutions	187
	4.3.2 Existence of Extremal Solutions	198
	4.3.3 Non-Well-Ordered Lower and Upper Solutions	201
4.4	Monotone Iterative Techniques	209
	4.4.1 Well Ordered Lower and Upper Solutions	210
	4.4.2 Reversed Ordered Lower and Upper Solutions	213
References		218

4.1 INTRODUCTION

This chapter is devoted to point out the influence that the constant sign of the related Green's functions has in order to ensure the existence of solutions of the considered nonlinear boundary value problems. Such existence results will be derived from fixed point theorems of a suitable integral operator whose fixed points coincide with the solutions of the studied problem. Several very well-known fixed point theorems, together with more recent ones, will be presented at the second section of this chapter. Existence of solutions of suitable boundary value problems, derived from the given fixed point theorems in cones, will be obtained. Furthermore, existence, nonexistence, and multiplicity of solutions of a one parameter family of periodic boundary value problems, where singularities on the spatial variable are considered, are deduced by combining the Krasnosel'skiĭ's fixed point Theorem.

In Sections 4.3 and 4.4 we will center our analysis on the lower and upper solutions method and the monotone iterative techniques. As we will

Maximum Principles for the Hill's Equation.
DOI: http://dx.doi.org/10.1016/B978-0-12-804117-8.00004-7
Copyright © 2018 Elsevier Inc. All rights reserved.

see, the fixed point theory showed in the second section will be fundamental to ensure the existence of solutions of the studied problems.

4.2 FIXED POINT THEOREMS AND DEGREE THEORY

This section is devoted to present some sufficient conditions that ensure the existence of at least one fixed point of operators defined in abstract spaces. This theory is fundamental to prove the existence of solutions of differential equations and, in particular, of the Hill's equation. This is due to the fact that the solutions of a given boundary value problem coincide with the fixed points of related integral operators that have as kernel the related Green's function in each case.

To illustrate this assertion, we can consider, for instance, the nonlinear periodic boundary value problem (as we will see the arguments for any other kind of boundary condition are analogous):

$$u''(t) + a(t) u(t) = f(t, u(t), u'(t)), \quad t \in I, \quad u(0) = u(T), \ u'(0) = u'(T).$$
$$(4.1)$$

Now, we assume that the homogeneous linear equation

$$u''(t) + a(t) u(t) = 0, \quad t \in I, \quad u(0) = u(T), \ u'(0) = u'(T), \qquad (4.2)$$

has the trivial solution as its unique one.

In such a case, as we have noted in Chapter 3, there is a unique Green's function $G_P[a] : I \times I \to \mathbb{R}$ related to the linear equation, and it is characterized by the properties $(G1)$–$(G5)$ (see also [8]). That is, u is a solution of the linear problem (with $\sigma \in L^1(I)$)

$$u''(t) + a(t) u(t) = \sigma(t), \quad t \in [0, T], \quad u(0) = u(T), \ u'(0) = u'(T),$$

if and only if

$$u(t) = \int_0^T G_P[a](t, s) \, \sigma(s) \, ds.$$

Moreover, we will work with f a L^1-Carathéodory function, i.e.,

Definition 11. We say that $f : I \times \mathbb{R}^2 \to \mathbb{R}$ is a L^1-Carathéodory function if it satisfies the following properties:
  (i)   for a.e. $t \in I$, the function $(u, v) \in \mathbb{R}^2 \to f(t, u, v) \in \mathbb{R}$ is continuous,

(ii)   for every $(u, v) \in \mathbb{R}^2$, the function $t \in I \rightarrow f(t, u, v)$ is measurable,

(iii) for every $R > 0$, there exists a real valued function $h_R \in L^1(I)$ such that

$$|f(t, u, v)| \leq h_R(t) \tag{4.3}$$

for a.e. $t \in I$ and every $(u, v) \in \mathbb{R}^2$ satisfying $|u| \leq R$, $|v| \leq R$.

As a consequence of this definition, the equality in the differential equation (4.1) can only be fulfilled a.e. on I. Therefore we cannot ensure that the solutions we are looking for are in $C^2(I)$. So, we will find solutions of the problem (4.1) that belong to the Sobolev space

$$W^{2,1}(I) := \left\{ u \in C^1(I), \quad u' \text{ is absolutely continuous on } I \right\}.$$

As a direct consequence, we have that the solutions of the nonlinear problem (4.1) are just the fixed points of the operator

$$\mathcal{T} : C^1(I) \rightarrow C^1(I)$$

defined as

$$\mathcal{T} u(t) := \int_0^T G_P[a](t, s) f(s, u(s), u'(s)) \, ds.$$

Thus, in order to deduce the existence of solution of nonlinear boundary value problems, we must use suitable results that ensure that operators defined on abstract spaces have fixed points. In this section we will present both classical and recent results in this line. To warrant the existence of solutions of the considered problems, some of these fixed point results will be used along the next sections of this chapter. First, we will present the Leray-Schauder degree and, as a consequence, we will deduce some suitable fixed point results for operators defined on normed and Banach spaces.

4.2.1 Leray-Schauder Degree

In this subsection we present the main properties of the, so called, Leray-Schauder degree. This concept was introduced by J. Leray and J. Schauder in 1934 in [67] and has been widely developed for several authors, among which we highlight the contributions of J.L. Mawhin [70]. It has become one of the most useful theories to ensure the existence of solutions of boundary value problems of different types, including ordinary and partial differential and difference equations. The main results we will present here can be consulted in [2,68]. In the monograph of A. Granas and J. Dugundji [44] classical and recent results concerning this theory are compiled.

First, we introduce the concept of completely continuous operator as follows (in [68] these operators are named as compact):

Definition 12. Let E and F be two real normed spaces, and $M \subset E$. The mapping $\mathcal{T} : M \to F$ is called completely continuous if and only if it satisfies the two following properties:
1. $\mathcal{T}$ is continuous.
2. $\mathcal{T}$ maps bounded sets of M into relatively compact sets of F.

The Leray-Schauder degree will be defined over operators of the form $I - \mathcal{T} : X \to X$, with X a normed space, I the identity operator and $\mathcal{T}$ completely continuous. For any $S \subset X$, $K(S)$ denotes the set of completely continuous operators defined on S and $K_1(S) = \{\phi, \phi = I - \mathcal{T}, \mathcal{T} \in K(S)\}$ denotes the completely continuous perturbations of the identity in S. The following definition is a particular case of [68, Chapter 4, Definition 5.1.1, and Theorem 5.1.2].

Definition 13. Let X be a normed vector space, D a nonempty, bounded, and open subset of X, $\phi \in K_1(\overline{D})$, and $p \in X \backslash \overline{\phi(\partial D)}$. The topological Leray–Schauder degree of ϕ at p relative to D, is defined as the unique integer $d(\phi, D, p)$, that satisfies the following axioms:
(i)   If $p \in D$ then $d(I, D, p) = 1$.
(ii) If D_1 y D_2 are two disjoint, bounded and open subsets of D and $\phi(u) \neq p$ for all $u \in \overline{D} \backslash (D_1 \cup D_2)$, then

$$d(\phi, D, p) = d(\phi|_{\overline{D_1}}, D_1, p) + d(\phi|_{\overline{D_2}}, D_2, p).$$

(iii) Let $h : [0, 1] \to K_1(\overline{D})$ and $\theta : [0, 1] \to X$ be both continuous, then, provided

$$\theta(t) \notin \overline{h(t)(\partial D)} \text{ for all } t \in [0, 1],$$

$d(h(t), D, \theta(t))$ is independent of $t \in [0, 1]$.

Assuming that the conditions in Definition 13 are satisfied, it is not difficult to verify the following results. The proofs can be found in [68, Section 4.3].

Theorem 26. *If $p \in D$ then $d(I, D, p) = 1$ and if $p \notin \overline{D}$ then $d(I, D, p) = 0$.*

Next result is the most used one for ensuring the existence of fixed points of completely continuous operators (it is enough to take $\phi = I - T$ and $p = 0$).

Theorem 27. *Let $\phi \in K_1(\overline{D})$ and $d(\phi, D, p) \neq 0$ then there exists $u \in D$ such that $\phi(u) = p$.*

Moreover, we have the main properties of the Leray-Schauder degree.

Theorem 28. *Assume that $\phi \in K_1(\overline{D})$ and $p \notin \phi(\partial D)$. Then the two following properties hold:*

(1) (Domain decomposition) If D is the disjoint union of open sets D_i, $i \in \mathbb{N}$, then

$$d(\phi, D, p) = \sum_{i=1}^{\infty} d(\phi, D_i, p).$$

(2) (Excision) If $K \subset \overline{D}$ is closed and $p \notin \phi(K)$, then

$$d(\phi, D, p) = d(\phi, D \backslash K, p).$$

Theorem 29 (Invariance under homotopy)**.** *Let $h(t)$ be a homotopy of completely continuous operators on $\overline{D}$ such that $p \notin (I - h(t))(\partial D)$ for all $t \in [0, 1]$. Then $d(I - h(t), D, p)$ is independent of $t \in [0, 1]$.*

As a direct consequence of this result, we deduce the following ones:

Theorem 30. *Let $\phi, \psi \in K_1(\overline{D})$ and $\phi = \psi$ on ∂D. Then, if $p \notin \phi(\partial D)$, $d(\phi, D, p) = d(\psi, D, p)$.*

Theorem 31. *Assume that $\phi \in K_1(\overline{D})$, $p \notin \phi(\partial D)$ and $q \in X$. Then $d(\phi, D, p) = d(\phi - q, D, p - q)$.*

Theorem 32. *Assume that $\phi \in K_1(\overline{D})$. Then $d(\phi, D, p)$ is constant for all p in the same connected component of $X \backslash \phi(\partial D)$.*

Previous result allows us to introduce the concept of index related to a point as follows.

Definition 14. Assume that $\phi \in K_1(\overline{D})$, $p \notin \phi(\partial D)$. Let $u_0 \in D$ be such that $\phi(u_0) = p$ and $\phi(u) \neq p$ for all u in a ball in D, centered at u_0. The index $i(\phi, u_0, p)$ is defined as $d(\phi, V, p)$, where V is any neighborhood of u_0 in D such that $\phi(u) \neq p$ for all $u \in V \backslash \{u_0\}$. When $p = 0$ and $\phi = I - T$ the index will be denoted as $i(T, V)$.

4.2.2 Fixed Point Theorems

As a consequence of the degree theory, one can obtain the following fixed point results. See [68, Section 4.4] for details.

The first one was proved by J. Schauder in 1930 [91]. The proof follows as a consequence of the degree axioms introduced in Definition 13.

Theorem 33. *Let S be a bounded, closed, nonempty, convex subset of the normed space X. Let $\phi \in K(S)$ be such that $\phi(S) \subset S$. Then ϕ has a fixed point in S.*

Theorem 34. *Let S be a bounded, closed subset of a normed space X, with nonempty interior, and let $\phi \in K(S)$. If there exists $w \in int(S) = D$ such that for all $\lambda > 1$ and $u \in \partial D$*

$$\phi(u) - w \neq \lambda (u - w), \tag{4.4}$$

then ϕ has a fixed point.

Theorem 35. *Suppose that X is a Banach space. Let S be a closed, convex subset of X (not necessarily bounded), and ϕ be a continuous mapping of S into a compact subset of S. Then ϕ has a fixed point.*

Theorem 36. *Suppose that X is a normed space (possibly incomplete). Let S be a bounded, closed, convex subset of X, containing the origin 0 in its interior D. Let $H : [0, 1] \times S \to X$ be a homotopy of completely continuous operators such that $H(0, \partial S) \subset S$ and*

$$H(t, u) \neq u \quad \text{for all } t \in [0, 1) \text{ and } u \in \partial S.$$

Then $\phi = H(1, \cdot)$ has a fixed point in S.

Next result was due to Schaefer in 1955 [89].

Theorem 37. *Let X be a normed space and $\phi \in K(X)$. Suppose that the set*

$$S = \{u \in X, \quad \text{for which there is some } \lambda \in [0, 1) \text{ such that } u = \lambda \phi u\}$$

is bounded in X. Then ϕ has, at least, a fixed point in X.

It is important to point out that the results derived from degree theory are not true, in general, for continuous operators. Next example can be found in [82].

Example 11. Let $l^2 = \{(u_1, u_2, \ldots) : \sum_{j=1}^{\infty} u_j^2 < \infty\}$ with the norm $\| u \| = \sqrt{\sum_{j=1}^{\infty} u_j^2}$, B the unit ball and $\mathcal{T} : \overline{B} \to \overline{B}$ defined as

$$\mathcal{T}(u) = \left(\sqrt{1 - \| u \|^2}, u_1, u_2, \ldots\right).$$

We know that l^2 is a Hilbert space with this norm and, obviously, $\mathcal{T}$ is a continuous operator.

Assume that there exists a map d defined on the set of continuous operators on $\overline{B}$ that satisfies the three properties of Definition 13. If this is true, we could ensure, by means of the Schauder's fixed point Theorem (Theorem 33) the existence of a fixed point u of operator $\mathcal{T}$ on $\overline{B}$. But, since $\|\mathcal{T}(u)\| = 1$ for all $u \in \overline{B}$, if there is a fixed point $u_* \in \overline{B}$ of $\mathcal{T}$, then $\|u_*\| = \|\mathcal{T}(u_*)\| = 1$. Moreover

$$(u_1, u_2, \ldots) = u_* = \mathcal{T}(u_*) = (0, u_1, u_2, \ldots)$$

and, therefore, $u_* = 0$.

However $\mathcal{T}(0) = (1, 0, 0, \ldots) \neq (0, 0, \ldots)$ and we arrive to a contradiction.

Other example in this direction can be found in [68, Section 4.1].

In order to ensure the existence of fixed points of operators defined in cones, one of the main results used was given by M.A. Krasnosel'skiĭ in 1964 [59]. As we will see, the proof of such result comes from degree theory. We follow the proofs of [2, Chapter III].

Before doing this, we must introduce some notations and definitions.

Definition 15. Let X be a Banach space. A subset $K \subset X$ is a cone if it is closed, $K + K \subset K$, $\lambda K \subset K$ for all $\lambda \geq 0$, and $K \cap (-K) = \{0\}$.

A cone K yields a partial ordering in X given by $x \leq y$ if and only if $y - x \in K$. X is an ordered normed space if X is ordered by a cone.

The notation $x < y$ means $x \leq y$ and $y \neq x$.

The cone K is called normal with normal constant $c \geq 1$ if and only if $\|x\| \leq c\|y\|$ for all $x, y \in X$ with $0 \leq x \leq y$.

Whenever $\operatorname{int}(K) \neq \emptyset$ the symbol $x \ll y$ means $y - x \in \operatorname{int}(K)$ and the cone is said solid.

∂K denotes the boundary of K, $d(x, \partial K)$ is the distance of x to the boundary of K and $K_\rho = \{x \in K, \|x\| < \rho\}$.

Lemma 45. *[2, Lemma 12.1] Let X be a Banach space and $K \subset X$ a cone in X. Let $\mathcal{T} : \overline{K_\rho} \longrightarrow K$ be a completely continuous operator. Then the two following properties hold:*

(i) *If $\mathcal{T}(u) \neq \lambda u$ for all $\lambda \geq 1$ and $u \in K$ such that $\|u\| = \rho$, then $i(\mathcal{T}, K_\rho) = 1$.*

(ii) *If there exists $p \in K$, $p > 0$, such that $u - \mathcal{T}(u) \neq \lambda p$ for all $\lambda \geq 0$ and $u \in K$ such that $\|u\| = \rho$, then $i(\mathcal{T}, K_\rho) = 0$.*

Proof. (i) Defining $h : [0, 1] \times \overline{K_\rho} \to K$ as $h(t, u) := t\,\mathcal{T}(u)$, it is obvious that $h(t, \cdot)$ is a completely continuous operator for any $t \in [0, 1]$. So, from Theorems 26 and 29, we have that

$$i(\mathcal{T}, K_\rho) = i(0, K_\rho) = 1.$$

(ii) Let $\mu := \sup\{\| \mathcal{T}(u) \|,\ u \in K_\rho\}$ and $\lambda_1 > (\rho + \mu)/\|p\|$. Then, by means of Theorem 29, we have

$$i(\mathcal{T}, K_\rho) = i(\mathcal{T} + \lambda_1 p, K_\rho).$$

If $i(\mathcal{T}, K_\rho) \neq 0$, then there is $u \in K_\rho$ such that $u = \mathcal{T}(u) + \lambda_1 p$. Thus

$$\|u\| \geq \lambda_1 \|p\| - \|\mathcal{T}(u)\| \geq \lambda_1 \|p\| - \mu > \rho$$

and we arrive at a contradiction.   □

As a direct consequence of the previous result, we obtain the following one.

Theorem 38. *Let X be a Banach space and $K \subset X$ a cone in X. Let $\mathcal{T} : \overline{K_\rho} \longrightarrow K$ be a completely continuous operator such that $\mathcal{T}(u) \neq \lambda u$ for all $\lambda \geq 1$ and $u \in K$ such that $\|u\| = \rho$. Then $\mathcal{T}$ has a fixed point in K_ρ.*

By means of Lemma 13 and Theorem 27, it is possible to get a sharp localization of the previously obtained fixed point.

Theorem 39. *Let X be a Banach space and $K \subset X$ a cone in X. Let $\mathcal{T} : \overline{K_\rho} \longrightarrow K$ be a completely continuous operator and $0 < \sigma, \tau \leq \rho$, $\sigma \neq \tau$. Assume that*

(i) *$\mathcal{T}(u) \neq \lambda u$ for all $\lambda \geq 1$ and $u \in K$ such that $\|u\| = \sigma$.*

(ii) *There is $p \in K$, $p > 0$, such that $u - \mathcal{T}(u) \neq \lambda p$ for all $\lambda \geq 0$ and $u \in K$ such that $\|u\| = \tau$.*

Then operator $\mathcal{T}$ has a fixed point u such that

$$\min\{\sigma, \tau\} < \|u\| < \max\{\sigma, \tau\}.$$

Proof. Let $\rho_0 := \min\{\sigma, \tau\}$ and $\rho_1 := \max\{\sigma, \tau\}$.

By Definition 13 (*ii*), we have that

$$i(\mathcal{T}, K_{\rho_1} \setminus \overline{K_{\rho_0}}) = i(\mathcal{T}, K_{\rho_1}) - i(\mathcal{T}, K_{\rho_0}).$$

Now, using Lemma 45, we have

$$\begin{cases} i(\mathcal{T}, K_{\rho_1} \setminus \overline{K_{\rho_0}}) = 1 & \text{if} \quad \rho_1 = \sigma, \\ i(\mathcal{T}, K_{\rho_1} \setminus \overline{K_{\rho_0}}) = -1 & \text{if} \quad \rho_1 = \tau. \end{cases}$$

In both cases $i(\mathcal{T}, K_{\rho_1} \setminus \overline{K_{\rho_0}}) \neq 0$ and, using Theorem 27, we conclude that there is a fixed point of operator $\mathcal{T}$ on $K_{\rho_1} \setminus \overline{K_{\rho_0}}$ and the result holds.
$\square$

As a direct consequence of previous results one can deduce the celebrated theorems of M.A. Krasnosel'skiĭ, of compression/expansion operators in cones [58].

Theorem 40 (Krasnosel'skiĭ's fixed point Theorem of compression in cones). *Let X be a Banach space and $K \subset X$ a cone in X. Let $\mathcal{T} : \overline{K_\rho} \longrightarrow K$ be a completely continuous operator and $0 < \sigma < \rho$. Suppose that $\mathcal{T}(u) \not\geq u$ for all $u \in K$ such that $\|u\| = \rho$, and $\mathcal{T}(u) \not\leq u$ for all $u \in K$ such that $\|u\| = \sigma$. Then $\mathcal{T}$ has a fixed point u such that $\sigma < \|u\| < \rho$.*

Theorem 41 (Krasnosel'skiĭ's fixed point Theorem of expansion in cones). *Let X be a Banach space and $K \subset X$ a cone in X. Let $\mathcal{T} : \overline{K_\rho} \longrightarrow K$ be a completely continuous operator and $0 < \sigma < \rho$. Suppose that $\mathcal{T}(u) \not\leq u$ for all $u \in K$ such that $\|u\| = \rho$, and $\mathcal{T}(u) \not\geq u$ for all $u \in K$ such that $\|u\| = \sigma$. Then $\mathcal{T}$ has a fixed point u such that $\sigma < \|u\| < \rho$.*

Despite the previous results, the most used fixed point theorem by M.A. Krasnosel'skiĭ was proved in 1964 [59] and it is as follows.

Theorem 42. *Let X be a Banach space and $K \subset X$ a cone in X. Let Ω_1 and Ω_2 be two open and bounded subsets of X, with $0 \in \Omega_1 \subset \overline{\Omega}_1 \subset \Omega_2$ and let $\mathcal{T} : K \cap (\overline{\Omega}_2 \setminus \Omega_1) \to K$ be a completely continuous operator satisfying one of the two following properties:*
(i) $\|\mathcal{T}u\| \leq \|u\|$ *for all* $u \in K \cap \partial\Omega_1$ *and* $\|\mathcal{T}u\| \geq \|u\|$ *for all* $u \in K \cap \partial\Omega_2$.
(ii) $\|\mathcal{T}u\| \geq \|u\|$ *for all* $u \in K \cap \partial\Omega_1$ *and* $\|\mathcal{T}u\| \leq \|u\|$ *for all* $u \in K \cap \partial\Omega_2$.
Then $\mathcal{T}$ has at least a fixed point on $K \cap (\overline{\Omega}_2 \setminus \Omega_1)$.

W. Feng and G. Zhang obtained in [38] a fixed point theorem, that is a consequence of the previous one and, in a similar direction, provides some location of the obtained fixed point. To this end we consider, in a Banach space X, ordered by a cone K, the ordered interval

$$[x, y] = \{z \in X : x \leq z \leq y\},$$

and define the subcone $\mathcal{K}_{u_0}$ as

$$\mathcal{K}_{u_0} = \{u \in K, u \geq \|u\| u_0\},$$

where $u_0 \in K$ is such that $\|u_0\| \leq 1$.

The obtained result is the following one.

Theorem 43 ([38, Theorem 2.1]). *Let X be an ordered Banach space with the order cone K. Let $0 \leq u_0 \leq \varphi$ be such that $\|u_0\| \leq 1$ and $\|\varphi\| = 1$, satisfying the following condition:*

$$if \quad u \in K, \ \|u\| \leq 1 \quad then \quad u \leq \varphi. \tag{4.5}$$

Then, if there exist two positive numbers $0 < c < b$ such that $\mathcal{T} : \mathcal{K}_{u_0} \cap (\overline{K}_b \setminus K_c) \to \mathcal{K}_{u_0}$ is a completely continuous operator and the conditions

$$\|\mathcal{T}u\|_{u \in [c\,u_0, c\,\varphi]} \leq c \quad and \quad \|\mathcal{T}u\|_{u \in [b\,u_0, b\,\varphi]} \geq b$$

or

$$\|\mathcal{T}u\|_{u \in [c\,u_0, c\,\varphi]} \geq c \quad and \quad \|\mathcal{T}u\|_{u \in [b\,u_0, b\,\varphi]} \leq b$$

are fulfilled, then the operator $\mathcal{T}$ has, at least, a fixed point $u \in [c\,u_0, b\,\varphi]$.

Finally, it is important to point out that, by means of the index theory properties, it is possible to ensure the existence of multiple fixed points of the considered operator. One of the most used ones is the, so-called, "three fixed points Theorem" due to H. Amann [1] (in that reference for strict set contraction operators).

Theorem 44. *Let X be a closed, bounded, convex subset of a Banach space and let X_1, X_2 be disjoint, closed, convex subsets of X. Let $\mathcal{T} : X \to X$ be a completely continuous operator and suppose that there exist open subsets O_1, O_2 of X with $O_i \subset X_i$, $i = 1, 2$. Moreover suppose that $T(X_i) \subset X_i$ and that $\mathcal{T}$ has no fixed points on $X_i \setminus O_i$, $i = 1, 2$. Then $\mathcal{T}$ has at least three distinct fixed points u, u_1, u_2 with $u_i \in X_i$, $i = 1, 2$, and $u \in X \setminus (X_1 \bigcup X_2)$.*

This kind of operators, defined in cones, are usually constructed with the aim of finding constant sign solutions (in the whole interval of definition or in some subinterval) of problems of the type (4.1) with nonresonant linear part.

As we will see in the sequel, if we are looking for positive solutions on the interval I, and the Green's function $G[a]$ related to the linear part of the considered equation is strictly positive on $I \times I$, the most used cone is the following one:

$$K = \left\{ u \in X, \ \min_{t \in I} u(t) \geq \frac{m}{M} \|u\| \right\},$$

where

$$0 < \min_{(t,s) \in I \times I} G[a](t, s) \equiv m < M \equiv \max_{(t,s) \in I \times I} G[a](t, s).$$

In some situations, as with the Dirichlet or the mixed boundary conditions, such property can never be fulfilled. In such a case, if the Green's function has constant sign, but it vanishes at some points of its square of definition, and if it satisfies some properties of the type:

There are two continuous functions g and Φ, strictly positive on $(0, T)$, such that

$$g(t)\, \Phi(s) \leq G(t, s) \leq \Phi(s), \quad \text{for all } (t, s) \in I \times I,$$

the cone can be defined, in this case, as

$$K = \{ u \in X, \ u(t) \geq g(t)\, \|u\| \text{ for all } t \in I \}.$$

Of course, these two types of cones are only an example of the ones that could be defined. In each particular case, it is necessary to define a cone according to the properties of the given Green's function. In next subsection some cones will be presented to deduce the existence of constant sign solutions of boundary value problems.

A combination of a small generalization of the Krasnosel'skiĭ's Fixed Point Theorem (Theorem 42) has been used in [11] to deduce existence, nonexistence, and multiplicity of solutions of a one parameter family of nonlinear equations coupled to periodic boundary conditions. Such results are compiled in Subsection 4.2.7.

Theorem 44 will be used in Section 4.3.3 in the framework of the lower and upper solutions method. It will allow us to ensure the existence of at least three solutions of nonlinear problems coupled with Dirichlet boundary conditions.

4.2.2.1 Application to Nonlinear Boundary Value Problems

In this subsection we will apply the previous fixed point theorems to deduce the existence of constant sign solutions of suitable second order nonlinear boundary value problems. In a first moment, we will consider the periodic problem

$$u''(t) + a(t) u(t) = f(t, u(t)) \text{ a.e. } t \in I, \ u(0) = u(T), \ u'(0) = u'(T), \quad (4.6)$$

where $a \in L^p(0, T)$, $1 \leq p \leq \infty$ and $f : I \times \mathbb{R} \to \mathbb{R}$ is an L^1-Carathéodory function.

In this case, we assume that the linear part of the equation is nonresonant (which excludes the case $a \equiv 0$) and the related Green's function has constant sign on its square of definition. A sub-linear behavior of f at $x = 0$ together with a super-linear one at $x = \infty$ (or vice-versa), is assumed. As we will point out, the used arguments are on the basis of the Krasnosel'skiĭ's fixed point Theorem (Theorem 42) and some of them can be adapted to other situations, which include different boundary conditions and the negativeness of the Green's function.

To be concise, we look for positive solutions of problem (4.6) by assuming the following hypothesis on the linear part:

($a0$)   The Hill's equation $u''(t) + a(t) u(t) = 0$ coupled with the corresponding boundary conditions is nonresonant (i.e., its unique solution is the trivial one) and the related Green's function is strictly positive on $I \times I$.

In Chapter 3 it has been done an exhaustive study on the potential a that ensures that such property is fulfilled by the corresponding Green's function. We mention that P. Torres in [94] proved some existence results for the periodic problem (4.6) under weaker assumptions on the nonlinear part of the equation.

In this case, by denoting

$$0 < \min_{(t,s) \in I \times I} G_P[a](t, s) \equiv m < M \equiv \max_{(t,s) \in I \times I} G_P[a](t, s),$$

we will work on the following cone:

$$K = \left\{ u \in C(I), \ \min_{t \in I} u(t) \geq \frac{m}{M} \|u\|_\infty \right\},$$

which is normal with $c = 1$ and has nonempty interior.

Let "$\preceq$" be the order induced in $C(I)$ by the cone K, i.e.,

$$x \preceq y \Longleftrightarrow \min_{t \in I}(y(t) - x(t)) \geq \frac{m}{M}\|y - x\|_\infty.$$

Moreover we denote:

$$f_0 := \lim_{x \to 0^+}\left\{\inf_{t \in I}\text{ ess}\,\frac{f(t,x)}{x}\right\}, \quad f^0 := \lim_{x \to 0^+}\left\{\sup_{t \in I}\text{ ess}\,\frac{f(t,x)}{x}\right\}$$

and

$$f_\infty := \lim_{x \to +\infty}\left\{\inf_{t \in I}\text{ ess}\,\frac{f(t,x)}{x}\right\}, \quad f^\infty := \lim_{x \to \infty}\left\{\sup_{t \in I}\text{ ess}\,\frac{f(t,x)}{x}\right\}.$$

Theorem 45. *Suppose that (a0) and the following assumptions hold:*
(f0) $f(t,x) \geq 0$ *for a.e.* $t \in I$ *and all* $x \geq 0$,
(f1) $f_0 = \infty$,
(f2) $f^\infty = 0$.
Then problem (4.6) has, at least, a positive solution.

Proof. As we have noticed in previous sections, the solutions of problem (4.6) are given as the fixed points of operator $T : C(I) \to C(I)$ defined as

$$Tu(t) = \int_0^T G_P[a](t,s)f(s,u(s))\,ds \text{ for all } t \in I, \tag{4.7}$$

where $G_P[a](t,s)$ is the corresponding Green's function.

In order to deduce the existence of a solution of the considered problem, we will ensure the existence of a fixed point of operator T on the cone K, by means of Krasnosel'skiĭ's fixed point Theorem (Theorem 42).

It is not difficult to verify that $T : K \to K$ is a completely continuous operator.

Let's see now that $T(K) \subset K$:

By (f0) we compute

$$\begin{aligned}
\min_{t \in I} Tu(t) &= \min_{t \in I}\left\{\int_0^T G_P[a](t,s)f(s,u(s))\,ds\right\} \geq \int_0^T mf(s,u(s))\,ds \\
&= \frac{m}{M}\int_0^T Mf(s,u(s))\,ds \geq \frac{m}{M}\max_{t \in I}\left\{\int_0^T G_P[a](t,s)f(s,u(s))\,ds\right\} \\
&= \frac{m}{M}\|Tu\|_\infty,
\end{aligned}$$

and then $Tu \in K$ for all $u \in K$.

Since $f_0 = \infty$, there exists a constant $\rho_1 > 0$ such that $f(t, u) \geq \delta_1 u$ for a.e. $t \in I$ and all $0 < u \leq \rho_1$, where $\delta_1 > 0$ satisfies

$$\delta_1 \frac{m}{M} \max_{t \in I} \left\{ \int_0^T G_P[a](t, s) \, ds \right\} \geq 1. \tag{4.8}$$

Take $u \in K$, such that $\|u\|_\infty = \rho_1$. So, we have

$$
\begin{aligned}
\|\mathcal{T} u\|_\infty &= \max_{t \in I} \left\{ \int_0^T G_P[a](t, s) \, f(s, u(s)) \, ds \right\} \\
&\geq \delta_1 \max_{t \in I} \left\{ \int_0^T G_P[a](t, s) \, u(s) \, ds \right\} \\
&\geq \delta_1 \|u\|_\infty \frac{m}{M} \max_{t \in I} \left\{ \int_0^T G_P[a](t, s) \, ds \right\} \geq \|u\|_\infty. \quad (4.9)
\end{aligned}
$$

Let $\delta_2 > 0$ be such that

$$\delta_2 \max_{t \in I} \left\{ \int_0^T G_P[a](t, s) \, ds \right\} \leq 1. \tag{4.10}$$

Using now that $f^\infty = 0$, there exists a constant $\rho_2 > \rho_1 > 0$ such that $f(t, u) \leq \delta_2 u$ for a.e. $t \in I$ and all $u \geq \rho_2$.

Let now $u \in K$ be such that $\|u\|_\infty = M \rho_2 / m$. As a consequence, since $u \in K$, we have that

$$u(t) \geq \frac{m}{M} \|u\|_\infty = \rho_2, \quad \text{for all } t \in I.$$

So, we deduce the following inequalities:

$$
\begin{aligned}
\|\mathcal{T} u\|_\infty &= \max_{t \in I} \left\{ \int_0^T G_P[a](t, s) \, f(s, u(s)) \, ds \right\} \\
&\leq \delta_2 \max_{t \in I} \left\{ \int_0^T G_P[a](t, s) \, u(s) \, ds \right\} \\
&\leq \delta_2 \|u\|_\infty \max_{t \in I} \left\{ \int_0^T G_P[a](t, s) \, ds \right\} \leq \|u\|_\infty. \quad (4.11)
\end{aligned}
$$

Thus, by Theorem 42, (ii), we can conclude that the periodic problem (4.6) has, at least, a positive solution. $\qquad\square$

Krasnosel'skiĭ's fixed point Theorem (Theorem 42) can also be used to deduce a dual result as follows.

Theorem 46. *Suppose that* (a0), (f0) *and the following assumptions hold:*

(f1)′ $f^0 = 0$

(f2)′ $f_\infty = \infty$

Then problem (4.6) has, at least, a positive solution.

Proof. Consider now the same operator T and cone K as in the proof of previous theorem.

Let $\delta_2 > 0$ be given as in Eq. (4.10). Since $f^0 = 0$, we have that there exists a constant $r_1 > 0$ such that $f(t, u) \leq \delta_2 u$ for a.e. $t \in I$ and all $0 \leq u \leq r_1$. Then, we have that inequality (4.11) is fulfilled for any $u \in K$ such that $\|u\|_\infty = r_1$.

Consider now $\delta_1 > 0$ satisfying (4.8). The fact that $f_\infty = \infty$ says us that there exists a constant $r_2 > r_1 > 0$ such that $f(t, u) \geq \delta_1 u$ for a.e. $t \in I$ and all $u \geq r_2$.

So, if $u \in K$ is such that $\|u\|_\infty = \frac{M}{m} r_2$, as in the proof of Theorem 45, we have that $u(t) \geq r_2$ for all $t \in I$ and inequality (4.9) is fulfilled.

Therefore, the result holds from Theorem 42, (i). $\qquad\square$

Remark 31. It is clear that the same result is fulfilled for Neumann problem. The conditions that ensure the positiveness of the related Green's function for this situation are also showed in Chapter 3. Both existence results are also valid for any other boundary conditions that ensure the positiveness of the Green's function.

When the related Green's function is strictly negative on $I \times I$, i.e., it is satisfied:

(b0)   The Hill's equation $u''(t) + a(t) u(t) = 0$ coupled with the corresponding boundary conditions is nonresonant (i.e., its unique solution is the trivial one) and the related Green's function satisfies that it is strictly negative on $I \times I$,

one can obtain dual results by changing the sign on both sides of the considered equation. So the conditions imposed to f in the two previous theorems must be assumed on $-f$. For this reason, in this case, the usual way to rewrite problem (4.6) is as follows

$$u''(t) + a(t) u(t) + f(t, u(t)) = 0 \text{ a.e. } t \in I, \quad u(0) = u(T), \quad u'(0) = u'(T),$$

$$(4.12)$$

where $a \in L^p(0, T)$, $1 \leq p \leq \infty$ and $f : I \times \mathbb{R} \to \mathbb{R}$ is a L^1-Carathéodory function.

The results are the following ones.

Theorem 47. *Suppose that conditions (b0) and (f0) are fulfilled. Moreover assume that one of the following situations holds:*
   (i)   (f1) and (f2),
   (ii) (f1)′ and (f2)′.
Then, problem (4.12) has, at least, a positive solution.

In some situations, as with the Dirichlet or the mixed boundary conditions, we have that the Green's function can never be positive on $I \times I$. In Chapter 3, it has been proved that it is negative on the interior of $I \times I$ whenever the first eigenvalue of the considered problem is positive. However, by construction of the Green's function, in particular from condition (G5), we know that it always vanishes at some points of the boundary of its square of definition and, as a direct consequence, condition (b0) is never fulfilled. So, Theorem 47 cannot be applied to this situation.

When this type of boundary conditions are considered, the Green's function may satisfy the following property
(b0)′   The Hill's equation $u''(t) + a(t)u(t) = 0$ coupled with the corresponding boundary conditions is nonresonant (i.e., its unique solution is the trivial one) and the related Green's function satisfies that it is strictly negative on $(0, T) \times (0, T)$.

To study the Dirichlet boundary conditions, we use the following property which is a direct consequence of [24, Remark 2.7] and improves [30, Chapter 3, Section 6].

Lemma 46. *Assume that condition (b0)′ holds for the Dirichlet conditions $u(0) = u(T) = 0$. Then the related Green's function $G_D[a]$ satisfies the following property:*
(b1)   *There exist $\Phi \in C(I)$, $\Phi > 0$ on $(0, T)$, and a positive constant $K > 0$, satisfying*

$$K s(s - T) \leq G_D[a](t, s) \leq \Phi(t) s(s - T), \quad \text{for all } (t, s) \in I \times I.$$

So, to work with the Dirichlet problem

$$u''(t) + a(t)u(t) + f(t, u(t)) = 0 \text{ a.e. } t \in I, \ u(0) = u(T) = 0, \quad (4.13)$$

where $a \in L^p(0, T)$, $1 \leq p \leq \infty$ and $f : I \times \mathbb{R} \to \mathbb{R}$ is an L^1-Carathéodory function, we consider the Banach space $X := C(I)$ with the supremum norm $\|\cdot\|_\infty$, ordered by the usual cone $\mathcal{K}$ of the continuous and nonnegative functions on I.

The existence result for problem (4.13) will be deduced, in this case, from Theorem 43. The idea follows similar steps to the ones given in [15]. To this aim, we define

$$u_0(t) = \frac{\Phi(t)}{K}, \quad t \in I,$$

which obviously satisfies that $u_0 \in X$ and $\|u_0\|_\infty \leq 1$.

So, we consider the following subcone:

$$\mathcal{K}_{u_0} = \left\{ u \in \mathcal{K}, \ u(t) \geq \frac{\Phi(t)}{K} \|u\|_\infty \text{ for all } t \in I \right\}.$$

The first existence result reads as follows.

Theorem 48. *Suppose that (b0)', (f0), (f1), and (f2) are fulfilled. Then problem (4.13) has, at least, a positive solution on (0, T).*

Proof. First, we note that the solutions of problem (4.13) are given as the fixed points of operator $\mathcal{T} : C(I) \to C(I)$ given by

$$\mathcal{T}u(t) = -\int_0^T G_D[a](t, s) f(s, u(s)) \, ds \quad \text{for all } t \in I,$$

where $G_D[a]$ is the corresponding Green's function.

It is immediate to verify that operator $\mathcal{T}$ is completely continuous.

From the regularity and non-negativeness of the functions $-G_D[a]$ and f on their domains of definition, we have that if $u \in \mathcal{K}_{u_0}$, then $\mathcal{T}u \in C(I)$ and $\mathcal{T}u(t) \geq 0$ for all $t \in I$.

Let us see that $\mathcal{T}(\mathcal{K}_{u_0}) \subset \mathcal{K}_{u_0}$.

Take $u \in \mathcal{K}_{u_0}$, then, for all $t \in I$, by using Lemma 46 the following inequalities are satisfied

$$\mathcal{T}u(t) = -\int_0^T G_D[a](t, s) f(s, u(s)) \, ds \geq \Phi(t) \int_0^T s(T - s) f(s, u(s)) \, ds$$

$$\geq \frac{\Phi(t)}{K} \int_0^T \max_{t \in I} \{-G_D[a](t, s)\} f(s, u(s)) \, ds$$

$$\geq \frac{\Phi(t)}{K} \max_{t \in I} \left\{ \int_0^T -G_D[a](t, s) f(s, u(s)) \, ds \right\} = \frac{\Phi(t)}{K} \|\mathcal{T}u\|_\infty.$$

Let $\varphi(t) = 1$ for all $t \in I$, it is clear that $0 \leq u_0 \leq \varphi$, $\|\varphi\|_\infty = 1$ and that condition (4.5) is fulfilled.

Since $f_0 = \infty$, there exists a constant $\rho_1 > 0$ such that $f(t, u) \geq \delta_1 u$ for a.e. $t \in I$ and all $0 \leq u \leq \rho_1$, where $\delta_1 > 0$ satisfies

$$\frac{\delta_1}{K} \max_{t \in I} \left\{ -\int_0^T \Phi(s)\, G_D[a](t, s)\, ds \right\} \geq 1. \tag{4.14}$$

Take $c = \rho_1$, then for $u \in [c u_0, c \varphi]$, from expression (4.14), we deduce the following inequalities

$$\|T u\|_\infty = \max_{t \in I} \left\{ -\int_0^T G_D[a](t, s) f(s, u(s))\, ds \right\}$$

$$\geq \delta_1 \max_{t \in I} \left\{ -\int_0^T G_D[a](t, s)\, u(s)\, ds \right\}$$

$$\geq \delta_1 \frac{c}{K} \max_{t \in I} \left\{ -\int_0^T \Phi(s)\, G_D[a](t, s)\, ds \right\} \geq c.$$

Moreover, from the continuity on the second variable of the function f, we can define the following function:

$$\tilde{f}(t, u) = \max_{z \in [0, u]} \left\{ f(t, z) \right\}, \quad t \in I, \ u \in \mathbb{R}.$$

Clearly, $\tilde{f}(t, \cdot)$ is a monotone non-decreasing function on $[0, \infty)$ for any $t \in I$ fixed.

Moreover, since $f^\infty = 0$, it is obvious (see [96]) that

$$\lim_{u \to \infty} \left\{ \sup_{t \in I} \mathrm{ess}\, \frac{\tilde{f}(t, u)}{u} \right\} = 0,$$

and then there exists a constant $\rho_2 > \rho_1 > 0$ such that $\tilde{f}(t, u) \leq \delta_2 u$ for a.e. $t \in I$ and all $u \geq \rho_2$, where $\delta_2 > 0$ satisfies

$$\delta_2 \max_{t \in I} \left\{ -\int_0^T G_D[a](t, s)\, ds \right\} \leq 1. \tag{4.15}$$

Let $b = \rho_2 \frac{K}{\tilde{M}}$, where $\tilde{M} = \max_{s \in I} \Phi(s) > 0$.

If we take $u \in [b u_0, b \varphi]$, then $b \geq \|u\|_\infty \geq \rho_2$ and from expression (4.15) we deduce the inequalities

$$\|T u\|_\infty \leq \max_{t \in I} \left\{ -\int_0^T G_D[a](t, s) \tilde{f}(s, u(s))\, ds \right\}$$

$$\leq \max_{t \in I} \left\{ -\int_0^T G_D[a](t,s) \tilde{f}(s, \|u\|_\infty) \, ds \right\}$$

$$\leq \delta_2 \|u\|_\infty \max_{t \in I} \left\{ -\int_0^T G_D[a](t,s) \, ds \right\} \leq b.$$

Thus, by the second part of Theorem 43, we conclude that problem (4.13) has, at least, a positive solution u such that $u \in [c\,u_0, b\varphi]$, and the result is proved. □

Theorem 49. *Suppose that* $(b0)'$, $(f0)$, $(f1)'$, *and* $(f2)'$ *are fulfilled. Then problem (4.13) has, at least, a positive solution on* $(0, T)$.

Proof. In this case, let $\delta_2 > 0$ be given as in Eq. (4.10). Since $f^0 = 0$, there exists a constant $r_1 > 0$ such that $f(t, u) \leq \delta_2 u$ for a.e. $t \in I$ and all $0 \leq u \leq r_1$. Take $b = r_1$, then for $u \in [b u_0, b \varphi]$ we deduce the following inequalities

$$\|T u\|_\infty = \max_{t \in I} \left\{ -\int_0^T G_D[a](t,s) f(s, u(s)) \, ds \right\}$$

$$\leq \delta_2 \|u\|_\infty \max_{t \in I} \left\{ -\int_0^T G_D[a](t,s) \, ds \right\} \leq b.$$

Now, from condition $(b1)$ we can ensure the existence of constants c_1, c_2, and δ_3 such that $0 < c_1 < c_2 < 1$ which satisfy

$$\frac{\delta_3}{K} \max_{t \in I} \left\{ -\int_{c_1}^{c_2} \Phi(s) \, G_D[a](t,s) \, ds \right\} \geq 1.$$

Since $f_\infty = \infty$, there exists a constant $r_2 > r_1 > 0$ such that $f(t, u) \geq \delta_3 u$ for all $u \geq r_2$.

Let $c = r_2 \frac{K}{\underline{M}}$, where $\underline{M} = \min_{s \in [c_1, c_2]} \Phi(s) > 0$. If we take $u \in [c u_0, c \varphi]$ we deduce the following inequalities

$$\|T u\|_\infty = \max_{t \in I} \left\{ -\int_0^T G_D[a](t,s) f(s, u(s)) \, ds \right\}$$

$$\geq \max_{t \in I} \left\{ -\int_{c_1}^{c_2} G_D[a](t,s) f(s, u(s)) \, ds \right\}$$

$$\geq \delta_3 \max_{t \in I} \left\{ -\int_{c_1}^{c_2} G_D[a](t,s) u(s) \, ds \right\}$$

$$\geq \delta_3 \frac{c}{K} \max_{t \in I} \left\{ -\int_{c_1}^{c_2} \Phi(s) \, G_D[a](t,s) \, ds \right\} \geq c.$$

Thus, by the second part of Theorem 43, we conclude that problem (4.13) has, at least, a positive solution u such that $u \in [c\,u_0, b\varphi]$. □

Remark 32. In order to improve our results in the two previous theorems, we can replace (respectively) the imposed conditions for the following ones:

(ĩ) $f_0 > \delta_1$ and $f^\infty < \delta_2$.

(ĩĩ) $f^0 < \delta_2$ and $f_\infty > \delta_3$.

It is not difficult to verify that, provided condition $(b0)'$ is fulfilled for the mixed boundary conditions, property $(b1)$ holds, with $-s$ instead of $s(s-T)$ for $u(0) = u'(T) = 0$, and with $s - T$ instead of $s(s-T)$ if $u'(0) = u(T) = 0$ are treated.

Several works have studied this kind of situations by looking for solutions that are not necessarily positive on the whole interval I (see [49–52] and references therein). In such a case, we can make some similar arguments by looking for positive solutions on a subinterval $[c, d] \subset I$ that avoid the extremes of I (or the corresponding one in case of mixed boundary conditions).

4.2.3 Extremal Fixed Points

Once the existence of at least one fixed point in a given set is established, it is important to know under which additional conditions on the operator $\mathcal{T}$ we can ensure the existence of extremal fixed points, i.e., are there any fixed points of the operator which are less than (or bigger than) any other fixed point on the given set? In the sequel we show the existence of such fixed points, proved by J.A. Cid in [29]. Before proving such a general result, we introduce some preliminary concepts and known properties.

Definition 16. We say that a subset Y of a partially ordered set (poset) X is upward directed if for each pair $y_1, y_2 \in Y$ there exists $y_3 \in Y$ such that $y_1 \leq y_3$ and $y_2 \leq y_3$. Analogously, Y is downward directed if for each pair $y_1, y_2 \in Y$ there exists $y_3 \in Y$ such that $y_3 \leq y_1$ and $y_3 \leq y_2$.

Definition 17. A poset X is a lattice if both $x_1 \vee x_2 := \sup\{x_1, x_2\}$ and $x_1 \wedge x_2 := \inf\{x_1, x_2\}$, exist for all $x_1, x_2 \in X$. Every totally ordered set is a lattice and every lattice is upward and downward directed. A lattice X is complete when each nonempty subset $B \subset X$ has supremum, denoted by $\bigvee B$, and infimum, denoted by $\bigwedge B$. In particular, every complete lattice has maximum and minimum.

In a normed space X, ordered by a cone K, we have that the intervals

$$(u] := \{z \in X \ : \ z \le u\} \quad \text{and} \quad [u) := \{z \in X \ : \ u \le z\}$$

are closed for all $u \in X$, because $(u] = u - K$ and $[u) = u + K$.

Definition 18. We say that $u^* \in D$ is the greatest fixed point of operator $\mathcal{T} : D \subset X \to X$ if u^* is a fixed point of $\mathcal{T}$ and $u \le u^*$ for any other fixed point $u \in D$. The smallest fixed point is defined similarly by reversing the inequality. When both, the least and the greatest fixed point of $\mathcal{T}$, exist we call them extremal fixed points.

The result that ensures the existence of extremal fixed points for operators defined in abstract spaces is the following one.

Theorem 50. *[29, Theorem 2.1] Let X be an ordered normed space, $D \subset X$ a nonempty, bounded, closed, and convex subset and $\mathcal{T} : D \to D$ a completely continuous operator. Then the set of fixed points of $\mathcal{T}$*

$$P = \{u \in D \ : \ \mathcal{T}u = u\},$$

is compact and nonempty.
   Moreover the following claims hold:
   (i) *$\mathcal{T}$ has a greatest (smallest) fixed point if and only if P is upward (downward) directed.*
   (ii) *If P is a lattice then P is a complete lattice.*

Proof. Schauder's fixed point Theorem (Theorem 33) ensures that P is nonempty. Moreover, P is closed, because $P = (\mathcal{T} - I)^{-1}(0)$, and since $\overline{\mathcal{T}(D)}$ is a compact set and $P = \mathcal{T}(P) \subset \overline{\mathcal{T}(D)}$, we deduce that P is also compact.

Proof of (i). If $\mathcal{T}$ has a greatest fixed point then, obviously, the set P is upward directed.

Conversely, suppose that P is upward directed. Then the following family of closed subsets of P

$$\mathcal{F}_1 = \{[u) \cap P \ : \ u \in P\}$$

has the finite intersection property.

Since P is compact we have that $\cap_{u \in P}([u) \cap P)$ contains a point u^*, which is the greatest fixed point of $\mathcal{T}$ because $u^* \in P$ and $u^* \in [u)$, i.e. $u^* \ge u$, for all $u \in P$.

By using dual arguments we prove that $\mathcal{T}$ has the smallest fixed point if and only if P is downward directed.

Proof of (ii). Suppose that P is a lattice and let $B \subset P$ be a nonempty subset. Since P is upward directed, we know, by claim (*i*), that $\mathcal{T}$ has a greatest fixed point u^*. Therefore the following family of closed subsets of P

$$\mathcal{F}_2 = \{[x, u] \cap P \; : \; x \in B, \; u \in P \text{ is an upper bound of } B\},$$

is nonempty, because $u^* \in P$ is an upper bound of B.

Moreover $\mathcal{F}_2$ has the finite intersection property because

$$\bigvee \{x_i \; : \; i = 1, \dots, n\} \in \bigcap_{i=1}^{n} ([x_i, u_i] \cap P),$$

for any $[x_i, u_i] \cap P \in \mathcal{F}_2$, $i \in \{1, \dots, n\}$.

Then, since P is compact, the intersection of all sets of the family $\mathcal{F}_2$ contains a point, which by construction is the supremum of B in P.

By using dual arguments we prove that there exists the infimum of B in P and thus P is a complete lattice. □

A list of different general conditions which imply that an upward directed set has maximum can be found in [47, Section 5].

An application of this result will be done in Section 4.3.2 to prove the existence of extremal solutions of the periodic boundary value problem (4.6).

4.2.4 Monotone Operators

As we have seen along previous subsections, by means of the degree theory it is possible to deduce the existence of fixed points of operators defined on abstract spaces. Moreover, by means of the finite intersection property, one can warrant the existence of the smallest and the greatest fixed point. The following classical result, which is a particular case of [1, Theorem 3] (in this reference it is proved for condensing operators), ensures that if the operator is monotone non-decreasing it is possible to obtain such extremal fixed points as the limit of recursive sequences.

Theorem 51. *Let X be an ordered real Banach space. Suppose that there exist $\alpha \leq \beta$ such that $\mathcal{T} \colon [\alpha, \beta] \subset X \to X$ is a completely continuous monotone non-decreasing operator with $\alpha \leq \mathcal{T}\alpha$ and $\mathcal{T}\beta \leq \beta$. Then $\mathcal{T}$ has a fixed point and the iterative sequence $\alpha_{n+1} = \mathcal{T}\alpha_n$, with $\alpha_0 = \alpha$, converges to the smallest fixed point* *of T in $[\alpha, \beta]$, and the sequence $\beta_{n+1} = T\beta_n$, with $\beta_0 = \beta$, converges to the greatest fixed point of T in $[\alpha, \beta]$.*

It is important to note that to apply previous result, it is fundamental to warrant that the operator is completely continuous. Despite this, in 1989 S. Heikkilä developed a method (see [46]) in which such regularity assumption is not necessary. In such result, the considered operator T can present some discontinuities. The used methods are on the basis of the so-called well ordered chain of T-iterations. One can consult the references [25,46,48] for details of the proofs together with very useful consequences related to monotone iterative techniques. In particular, by means of these techniques, monotone methods for boundary value problems with discontinuous right hand sides have been obtained in [17,18].

Now, we present two useful versions of the general fixed points results obtained under this theory. The first one is a direct adaptation of [48, Proposition 1.4.4].

Theorem 52. *Given an order interval $[\alpha, \beta] \subset AC(I)$ and $T : [\alpha, \beta] \to [\alpha, \beta]$ a monotone non-decreasing operator such that there is $M \in L^1(I)$ satisfying*

$$|(T\eta)'(t)| \leq M(t) \text{ for a.e. } t \in I \text{ and all } \eta \in [\alpha, \beta].$$

Then T has the smallest fixed point u_* and the greatest fixed point u^*. Moreover,

$$u_* = \min\{u \in [\alpha, \beta] \mid T u \leq u\} \quad and \quad u^* = \max\{u \in [\alpha, \beta] \mid u \leq T u\}.$$

Next result is proved in [18, Lemma 2.2] and follows from [48, Proposition 1.2.2].

Lemma 47. *Given an order interval $[\alpha, \beta] \subset C(I)$ and a mapping $T : [\alpha, \beta] \to [\alpha, \beta]$, assume that T is monotone non-decreasing, and that the sequence $(T v_n)$ has a pointwise limit in $C(I)$ whenever (v_n) is a monotone sequence in $[\alpha, \beta]$. Then T has the smallest fixed point u_* and the greatest fixed point u^*. Moreover,*

$$u_* = \min\{u \in [\alpha, \beta] \mid T u \leq u\} \quad and \quad u^* = \max\{u \in [\alpha, \beta] \mid u \leq T u\}.$$

The points α and β that satisfy the inequalities in previous results are known as lower and upper solutions. As we have seen, under monotonicity assumptions on the operator T, it is possible to ensure the existence of extremal fixed points between both of them. However, to apply Krasnosel'skiǐ's fixed point Theorem no monotonicity is required. In the rest of this section, we present some results, proved in [10], where the combination of these two theories is done.

In [85], by using the properties of the topological degree, H. Persson gave sufficient conditions for the existence of a non-zero fixed point for a monotone non-decreasing mapping $f : \mathbb{R}_+^m \to \mathbb{R}_+^m$. This result has applications in the study of economic models and, surprisingly enough, it results equivalent to Brouwer fixed point theorem through the use of the Knaster-Kuratowski-Mazurkiewicz Lemma (see [84]).

The following result is proved in [10] and gives sufficient conditions for a monotone non-decreasing operator defined on an ordered Banach space to have at least a positive non-zero fixed point.

Theorem 53. *Let X be a real Banach space, K a normal and solid cone, and $\mathcal{T} : K \to K$ a monotone non-decreasing and completely continuous operator. Define*

$$S = \{u \in K : \mathcal{T}u \le u\}$$

and suppose that
(i)   *There exists $\bar{u} \in S$ such that $\bar{u} \gg 0$.*
(ii)   *S is bounded.*
    Then there exists $u \in K$, $u \ne 0$, such that $u = \mathcal{T}u$.

Proof. Since $\bar{u} \in \text{int}(K)$ there exists $r > 0$ such that

$$\overline{B(\bar{u}, r)} := \{u \in X; \ \|u - \bar{u}\| \le r\} \subset K.$$

Now, if $u \in K$ with $\|u\| = r$ it is clear that $\bar{u} - u \in \overline{B(\bar{u}, r)} \subset K$ and therefore $u \le \bar{u}$.

Now we distinguish two cases:

Case (i) Firstly, suppose that there exists $u \in K$ with $\|u\| = r$ such that $\mathcal{T}u \ge u$. Let us define the sequence $u_0 = u$, $u_n = \mathcal{T}u_{n-1}$ for all $n \in \mathbb{N}$. Since $u \le \bar{u}$, $\bar{u} \in S$ and $\mathcal{T}$ is monotone non-decreasing we have that

$$0 < u \le u_n = \mathcal{T}^n u \le \mathcal{T}^n \bar{u} \le \bar{u} \quad \text{for all } n \in \mathbb{N}.$$

Then the normality of K implies that $\|u_n\| \le c\|\bar{u}\|$, that is, $\{u_n\}_{n=0}^\infty$ is bounded. Now, as $\mathcal{T}$ is a completely continuous operator, the sequence $\{\mathcal{T}u_n\}_{n=0}^\infty = \{u_n\}_{n=0}^\infty$ is relatively compact and therefore there exists a subsequence $\{u_{n_k}\}_{k=1}^\infty$ converging to a point u^*.

Notice that, since $\mathcal{T}$ is monotone non-decreasing, $u_{n_k} \leq u^*$ for each k. Thus for each $n \geq n_k$ we have $u_{n_k} \leq u_n \leq u^*$ and, from the normality of K, it follows that

$$\|u^* - u_n\| \leq c \|u^* - u_{n_k}\|,$$

which implies that the whole sequence $\{u_n\}_{n=0}^{\infty} \to u^*$.

Since $\mathcal{T}$ is continuous, we deduce that $u^* = \mathcal{T}u^*$ and, therefore, u^* is a fixed point of $\mathcal{T}$ such that $0 < u \leq u^* \leq \bar{u}$. In particular $u^* \in K \setminus \{0\}$.

Case (ii) On the contrary, suppose that $\mathcal{T}u \not\geq u$ for all $u \in K$ with $\|u\| = r$. Now, since S is bounded, there exists $R > r$ such that $\mathcal{T}u \not\leq u$ for all $u \in K$ with $\|u\| = R$. Thus Theorem 41 implies the existence of a non-zero fixed point also in this case. □

Remark 33. (I) Clearly condition (*ii*) can be replaced by the weaker one (*ii*)$_*$ There exists $R > 0$ such that $S \cap \{u \in K : \|u\| = R\} = \emptyset$.

(II) Theorem 53 combines the monotone iterative technique with the expansion fixed point theorem of Krasnosel'skiĭ. Of course, the classical monotone method, presented in Theorem 51, is also applicable under our assumptions, but it does not exclude the zero fixed point. Actually, if $\mathcal{T}(0) > 0$ and $S \neq \emptyset$, as we have noted in Theorem 52 and Lemma 47, the existence of a non-zero fixed point for the monotone non-decreasing operator $\mathcal{T}$ follows from the monotone method under much weaker assumptions than continuity and compactness (see the monograph [48]). Hence, the significant case for us is whenever the zero fixed point is already known, that is, when $\mathcal{T}(0) = 0$.

On the other hand, the main difference between Theorem 53 and Krasnosel'skiĭ's type fixed point theorems (see for instance [36,98] or the generalization given in [100]) is that it is possible to assume condition $\mathcal{T}(x) \not\geq x$ for all $x \in K$, $\|x\| = \rho$, but the other inequality can be replaced by the monotonicity of the operator. As we will see in Subsection 4.2.4.1, in applications to ordinary differential equations, Krasnosel'skiĭ's assumptions turn usually into conditions over the asymptotic behavior of the nonlinearity on 0 and $+\infty$, respectively. Theorem 53 shall allow us to impose only a prescribed asymptotic behavior on $+\infty$ but nothing is said about 0.

(III) Under the assumptions of Theorem 53 the monotonicity condition of $\mathcal{T} : K \to K$ can be improved by the following one: there exists a real constant $M \geq 0$ such that the operator $\mathcal{T}x + Mx$ is monotone non-decreasing on K. The proof of this fact relies on the equivalence between the fixed point equation $\mathcal{T}x = x$ and $\mathcal{A}x = x$, where

$$\mathcal{A}x \equiv \frac{1}{1+M}(\mathcal{T}+M)x.$$

Moreover, it is easy to show that $A : K \to K$ is a condensing operator (see [98, Example 11.7]) and $\mathcal{A}x \leq x$ if and only if $\mathcal{T}x \leq x$. (Notice that the combination of monotone iterative technique and Theorem 41 given in the proof of Theorem 53 is also valid for condensing maps.)

As a particular case of Theorem 53, with $N = \mathbb{R}^m$ and the cone $K = \mathbb{R}^m_+$, we obtain the following result.

Corollary 42. *([85, theorem 5]) Assume that $f : \mathbb{R}^m_+ \to \mathbb{R}^m_+$ is continuous and monotone non-decreasing. Let $S = \{u \in \mathbb{R}^m_+ : f(u) \leq u\}$. If S is bounded, and there is $u' \in S$, $u' \gg 0$, then there is $u \geq 0$, $u \neq 0$, such that $u = f(u)$.*

Theorem 53 has been improved by J.A. Cid, D. Franco, and F. Minhós in [28]. The given result is the following one.

Theorem 54. *[28, Theorem 2.3] Let X be a real Banach space, K a normal and solid cone, with normal constant $c \geq 1$, and $\mathcal{T} : K \to K$ a completely continuous operator.*
Define $S = \{x \in K : \mathcal{T}x \leq x\}$ and suppose that
(i)   There exist $\beta \in S$ and $\bar{R} > 0$ such that $B[\beta, \bar{R}] \subset K$, i.e., $\beta \in S \cap int(K)$, and that one of the two following conditions holds:
(ii) S is bounded;
(iii) There exists $r > 0$ such that $S \cap B[0, r] = \emptyset$.
If moreover $\mathcal{T}$ is monotone non-decreasing in the set

$$K_1 = \left\{ x \in K : \frac{\bar{R}}{c} \leq \|x\| \leq c\|\beta\| \right\},$$

where $c \geq 1$ is the normal constant of the cone, then there exists a non-trivial $u \in K$, such that $u = \mathcal{T}u$.

Proof. Since $B[\beta, \bar{R}] \subset K$ it is clear that if $u \in K$ with $\|u\| = \bar{R}$ then $u \leq \beta$.

First suppose that we can choose $\alpha \in K$ with $\|\alpha\| = \bar{R}$ and $\mathcal{T}\alpha \geq \alpha$. Since $\alpha \leq \beta$ and due to the normality of the cone K we have that $[\alpha, \beta] \subset K_1$ which implies that $\mathcal{T}$ is monotone non-decreasing on $[\alpha, \beta]$. Then we can apply the classical monotone iterative technique, developed in Theorem 51, to ensure the existence of the extremal fixed points of $\mathcal{T}$ on $[\alpha, \beta]$, which are, in particular, positive fixed points.

Now suppose that $\mathcal{T}u \not\geq u$ for all $u \in K$ with $\|u\| = \bar{R}$, i.e. that we can not choose α as in the above paragraph. If (ii) holds we can take $R > \bar{R}$ such that $\mathcal{T}u \not\leq u$ for all $u \in K$ with $\|u\| = R$ and if (iii) holds we pick $0 < R < \bar{R}$ satisfying the same condition. In both cases, by Theorem 42, there is a positive non-zero fixed point. $\qquad\square$

Remark 34. (1) The main improvement of Theorem 54 with respect to Theorem 53 is that the operator $\mathcal{T}$ is assumed to be monotone non-decreasing only on the set K_1 and not on the whole cone K.

(2) Theorem 54 mixes the assumptions of the monotone iterative technique with the conditions of cone expansions or compressions. In our opinion one of the most interesting features of the above theorem is that only conditions over the set S of the upper solutions are assumed and the existence of a positive well-ordered lower solution is not needed.

Previous result has been generalized to a wider situation by D. Franco, G. Infante, and J. Perán in [39]. The main novelty of this result with respect to Theorem 54 is that the set S is not required to be neither bounded nor bounded away from 0. The result is the following one.

Theorem 55. *[39, Theorem 3.1] Let X be a real Banach space, K a normal solid cone with normal constant $c \geq 1$ and $\mathcal{T}: K \to K$ a completely continuous operator. Assume that*
(I1)   there exist $\beta_1 \in K$, with $\mathcal{T}\beta_1 \leq \beta_1$, and $R_1 > 0$ such that $B[\beta_1, R_1] \subset K$,
(I2)   the map $\mathcal{T}$ is monotone non-decreasing in the set

$$\tilde{K} = \left\{ u \in K : \frac{R_1}{c} \leq \|u\| \leq c\|\beta_1\| \right\},$$

(I3)   there exists $r_1 > 0$, with $r_1 \neq R_1$, such that $\mathcal{T}u \not\leq u$ for all $u \in K$ with $\|u\| = r_1$.
Then the map $\mathcal{T}$ has, at least, a non-zero fixed point u_1 in K, that either belongs to $\tilde{K}$ or is such that

$$\min\{r_1, R_1\} < \|u_1\| < \max\{r_1, R_1\}.$$

Proof. Since $B[\beta_1, R_1] \subset K$ we have that if $u \in K$ is such that $\|u\| = R_1$, then $u \leq \beta_1$.

First suppose that we can choose $\alpha_1 \in K$, with $\|\alpha_1\| = R_1$ and $\mathcal{T}\alpha_1 \geq \alpha_1$. Since $\alpha_1 \leq \beta_1$ and due to the normality of the cone K, we have that

$[\alpha_1, \beta_1] \subset \tilde{K}$, which implies that operator $\mathcal{T}$ is monotone non-decreasing on $[\alpha_1, \beta_1]$. Then we can apply Theorem 51 to ensure the existence of extremal fixed points of $\mathcal{T}$ on $[\alpha_1, \beta_1]$, which are, in particular, non-trivial fixed points.

Now suppose that such an α_1 does not exist. Thus, $\mathcal{T}u \not\geq u$ for all $u \in K$ with $\|u\| = R_1$. Since $\mathcal{T}u \not\leq u$ for all $u \in K$ such that $\|u\| = r_1$, we deduce, by Theorem 42, the existence of a non-trivial fixed point u_1.   □

Remark 35. The completely continuity assumption in Theorem 55 can be relaxed. In fact the result is valid for condensing maps. This is due to the fact that Theorems 42 and 51 are valid for condensing maps (see, for example, [98]).

As a combination of this kind of fixed point theorems, multiplicity of solutions has been proved in [39, Theorem 3.4].

4.2.4.1 Existence of Solutions of Periodic Boundary Value Problems

In the sequel we will deduce the existence of positive solutions by using Theorem 53. The result is the following one, and it improves [10, Theorem 3.2].

Theorem 56. *Suppose that (a0), (f0), (f2)$'$, and the following assumptions hold:*

(f3)   There exists $\bar{u} \in W^{2,1}(I)$ with $\min\limits_{t\in I} \bar{u}(t) > 0$, $\bar{u}(0) = \bar{u}(T)$, $\bar{u}'(0) = \bar{u}'(T)$

and moreover

$$\bar{u}''(t) + a(t)\bar{u}(t) \geq f(t, \bar{u}(t)) \quad a.e.\ t \in I.$$

(f4) $f(t, \cdot)$ is monotone non-decreasing a.e. $t \in I$.
Then problem (4.6) has, at least, a positive solution.

Proof. As in the proof of Theorem 45, we have that the solutions are given as the fixed points of the completely continuous operator $\mathcal{T}$ defined in (4.7).

In this case, denoting again

$$m = \min_{t,s\in I} G_P[a](t,s) \quad \text{and} \quad M = \max_{t,s\in I} G_P[a](t,s),$$

for any $0 < \gamma < \frac{m}{M}$, we consider the following cone:

$$K_\gamma = \{u \in C(I),\ u(t) \geq \gamma \|u\|_\infty \text{ for all } t \in I\}.$$

As in the proof of Theorem 45, from condition $(f0)$ we deduce that $\mathcal{T}(K_\gamma) \subset K_\gamma$.

In order to apply Theorem 53, we will divide the proof into several steps.

First, we prove that $\mathcal{T} : K \to K$ is monotone non-decreasing. Using $(f4)$ and arguing as in the proof of Theorem 45, it is not difficult to verify that if $x \le y$ then

$$\min_{t \in I} \left(\mathcal{T} y(t) - \mathcal{T} x(t) \right) \ge \frac{m}{M} \| \mathcal{T} y - \mathcal{T} x \|_\infty \ge \gamma \| \mathcal{T} y - \mathcal{T} x \|_\infty.$$

Thus $\mathcal{T}$ is monotone non-decreasing with respect to the partial ordering "$\preceq$" induced by K_γ.

Let's see now that $\bar{u} \in S = \{ u \in K_\gamma : \mathcal{T} u \le u \}$ and $\bar{u} \gg 0$.

Since $\min_{t \in I} \bar{u}(t) > 0$ we can choose $0 < \gamma < \frac{m}{M} < 1$ small enough such that

$$\min_{t \in I} \bar{u}(t) > \gamma \| \bar{u} \|_\infty,$$

and then $\bar{u} \in \mathrm{int}(K_\gamma)$.

On the other hand, condition $(f3)$ ensures us the existence of a non-negative function $h \in L^1(I)$ such that

$$\bar{u}''(t) + a(t)\, \bar{u}(t) = f(t, \bar{u}(t)) + h(t) \quad \text{a.e. } t \in I,$$

which is equivalent to

$$\bar{u}(t) - \mathcal{T} \bar{u}(t) = \int_0^T G_P[a](t, s)\, h(s)\, ds.$$

Now, by similar computations to previous steps, we arrive at

$$\min_{t \in I} (\bar{u}(t) - \mathcal{T} \bar{u}(t)) \ge \gamma \| \bar{u} - \mathcal{T} \bar{u} \|_\infty,$$

which implies $\mathcal{T} \bar{u} \preceq \bar{u}$.

So, to finish the proof, we only need to verify that the set S is bounded. By $(f2)'$ there exists $r > 0$ such that

$$f(t, x) > \frac{x}{\gamma\, m\, T} \quad \text{for all } x > r \text{ and a.e. } t \in I.$$

Let $u \in K$ be such that $\min_{t \in I} u(t) > r$. Then

$$f(t, u(t)) > \frac{u(t)}{\gamma\, m\, T} \quad \text{for a.e. } t \in I,$$

and we obtain for all $t \in I$

$$
\begin{aligned}
\mathcal{T}u(t) &= \int_0^T G_P[a](t,s) f(s,u(s))\, ds > \int_0^T G_P[a](t,s) \frac{u(s)}{\gamma \, m \, T}\, ds \\
&\geq \int_0^T m \, \frac{\gamma \, \|u\|_\infty}{\gamma \, m \, T}\, ds = \|u\|_\infty \geq u(t),
\end{aligned}
$$

so $\mathcal{T}u \not\leq u$ and in consequence $u \notin S$.

Then, whenever $u \in S$ we have that $\min_{t \in I} u(t) \leq r$ and from

$$
r \geq \min_{t \in I} u(t) \geq \gamma \, \|u\|_\infty,
$$

it follows that $S \subset \overline{B\left(0, \frac{r}{\gamma}\right)}$.

Finally, as a consequence of all the previous steps and Theorem 53 it follows the existence of a non-trivial fixed point in K for operator $\mathcal{T}$, which is a positive solution of the periodic problem (4.6).   □

Remark 36. Notice that in Theorem 56 we have obtained the same result as in Theorem 41, by replacing the sub-linear growth condition at 0, $(f1)'$, by the existence of a lower solution $(f3)$, and the monotonicity assumption $(f4)$.

Remark 37. (1) Several results on the existence and multiplicity of positive solutions for different kinds of boundary value problems can be found in [57,69,94] and references therein.

(2) In condition $(f3)$ we assume the existence of a positive lower solution $\alpha = \bar{u}$ for Eq. (4.6). As it will be pointed out in Section 4.4, when the Green's function is positive on $I \times I$, as in the case studied here, the existence of a lower and an upper solution satisfying the reversed order, $\beta \leq \alpha$ on I, implies the existence of a solution between them (see [5,34]). Of course, in our case we could choose $\beta \equiv 0$ as an upper solution, but then the trivial solution is not excluded. In Theorem 56 the a-priori bound on the lower solution set S is the fundamental key to ensure the existence of a positive solution.

Remark 38. As in previous cases, it is clear that the same result is fulfilled for the Neumann problem and for any other boundary conditions that ensure the strict positiveness of the Green's function on $I \times I$.

Moreover, when the related Green's function is strictly negative on $I \times I$, i.e., condition $(b0)$ is satisfied, one can obtain a dual result for the periodic problem (4.12) as follows:

Theorem 57. *Suppose that conditions* $(b0)$, $(f0)$, $(f2)'$, $(f4)$ *are fulfilled and, moreover, the following assumption holds:*

$(f3)'$ *There exists* $\bar{u} \in W^{2,1}(I)$ *with* $\min_{t\in I} \bar{u}(t) > 0$, $\bar{u}(0) = \bar{u}(T)$, $\bar{u}'(0) = \bar{u}'(T)$
   and moreover

$$\bar{u}''(t) + a(t)\,\bar{u}(t) + f(t, \bar{u}(t)) \le 0 \quad a.e.\ t \in I.$$

Then, problem (4.12) has, at least, a positive solution.

Remark 39. In this case, Theorem 57 gives an existence result, which is analogous to Theorem 47, *(ii)*, by replacing the sub-linear growth condition at 0, $(f1)'$ by the existence of an upper solution $(f3)'$, and the monotonicity assumption $(f4)$.

4.2.5 Non-increasing Operators

In the sequel the results proved in [13] for monotone non-increasing operators are showed. Firstly, we present a result under the assumption of the existence of a lower solution in a solid cone. The proofs adapt the ideas developed in Theorems 51, 53 and 54 to non-increasing operators.

Theorem 58. *Let X be a real Banach space, K a solid cone and $\mathcal{T}: K \to K$ a completely continuous operator. Assume that*
$(T1)$ *there exist $\alpha_1 \in K$, with $\alpha_1 \le \mathcal{T}\alpha_1$, and $R_1 > 0$ such that $B[\alpha_1, R_1] \subset K$,*
$(T2)$ *the map $\mathcal{T}$ is monotone non-increasing in the set*

$$K_2 = \{u \in K : R_1 \le \|u\| \le \|\alpha_1\|\},$$

$(T3)$ *there exists $r_1 > 0$, with $r_1 \ne R_1$, such that $\mathcal{T}u \not\geq u$ for all $u \in K$ with $\|u\| = r_1$.*
Then the map $\mathcal{T}$ has, at least, a non-zero fixed point $u_1 \in K\backslash\{u \in K; \|u\| = r_1\}$, such that

$$\min\{r_1, R_1\} \le \|u_1\| \le \max\{r_1, R_1\}.$$

Proof. First, note that $0 < R_1 \le d(\alpha_1, \partial K) \le \|\alpha_1\|$. Since $B[\alpha_1, R_1] \subset K$, it is clear that if $u \in K$ with $\|u\| = R_1$, then $u \le \alpha_1$. Since $\mathcal{T}$ is monotone non-increasing in K_2 and u, $\alpha_1 \in K_2$, we have that

$$\mathcal{T}u \ge \mathcal{T}\alpha_1 \ge \alpha_1 \ge u \quad \text{for all } u \in K \text{ with } \|u\| = R_1.$$

If $\mathcal{T}u = u$, we have a fixed point such that $\|u\| = R_1$, on the contrary, we deduce that $\mathcal{T}u \not\leq u$ for all $\|u\| = R_1$, which together with (T3) implies, by

Theorems 40 and 41, that there exists a non-zero fixed point u_1 with the desired localization property.   □

Remark 40. Note that if $\mathcal{T}(\alpha_1) \neq \alpha_1$ in the proof of the previous result it is showed that the map $\mathcal{T}$ has, at least, a non-zero fixed point $u_1 \in K$, such that

$$\min\{r_1, R_1\} < \|u_1\| < \max\{r_1, R_1\}.$$

As a direct consequence of Theorem 58, we obtain the following "dual result" for a monotone non-increasing operator in a solid cone with an upper solution.

Corollary 43. *Let X be a real Banach space, K a solid cone and $\mathcal{T}: K \to K$ a completely continuous operator. Assume that condition (T3) is fulfilled together with*

(T4) *there exist $\beta_1 \in K$, with $\mathcal{T}\beta_1 \leq \beta_1$, and $R_1 > 0$ such that $B[\mathcal{T}\beta_1, R_1] \subset K$,*

(T5) *the map $\mathcal{T}$ is monotone non-increasing in the set*

$$K_3 = \{u \in K : R_1 \leq \|u\| \leq \max\{\|\mathcal{T}\beta_1\|, \|\beta_1\|\}\}.$$

Then the map $\mathcal{T}$ has, at least, a non-zero fixed point u_1 in $u_1 \in K \backslash \{u \in K; \|u\| = r_1\}$, and

$$\min\{r_1, R_1\} \leq \|u_1\| \leq \max\{r_1, R_1\}.$$

Proof. As in the proof of Theorem 58 we deduce that $\|\mathcal{T}\beta_1\| \geq R_1$.

If $\|\beta_1\| = \|\mathcal{T}\beta_1 - \beta_1 - \mathcal{T}\beta_1\| \leq R_1$, then $\mathcal{T}\beta_1 - \beta_1 \in B[\mathcal{T}\beta_1, R_1] \subset K$, which implies that $\mathcal{T}\beta_1 - \beta_1 \geq 0$, i.e. β_1 is a fixed point of $\mathcal{T}$. Obviously, if this is the case, $\|\beta_1\| = \|\mathcal{T}\beta_1\| = R_1$ and the result holds.

On the other hand, if $\|\beta_1\| > R_1$, we know that β_1, $\mathcal{T}\beta_1 \in K_3$ and, since $\mathcal{T}$ is monotone non-increasing in K_3 and $\mathcal{T}\beta_1$ belongs to the interior of K, we deduce that $0 \ll \mathcal{T}\beta_1 \leq \mathcal{T}(\mathcal{T}\beta_1)$, i.e. $\mathcal{T}\beta_1 \in K$ is a non-zero lower solution of the operator $\mathcal{T}$.

As a consequence, all the conditions of Theorem 58 are fulfilled and the result holds.   □

Now, we give a fixed point result for a monotone non-increasing operator under the assumption of the existence of an upper solution in a not necessarily solid cone, but that verifies an extra condition.

Theorem 59. *Let X be a real Banach space, K a cone that satisfies the following condition*

there exists $\sigma \geq 1$ such that for $x, y \in K$ with $\|x\| = \sigma \|y\|$ we have $x \geq y$,

$$(4.16)$$

and $\mathcal{T}: K \to K$ a completely continuous operator.

Assume that

(T6) there exists $\beta_1 \in K$, $\beta_1 \neq 0$ such that $\beta_1 \geq \mathcal{T}\beta_1$,

(T7) the map $\mathcal{T}$ is monotone non-increasing in the set

$$K_4 = \{u \in K : \|\beta_1\| \leq \|u\| \leq \sigma \|\beta_1\|\},$$

(T8) there exists $r_1 > 0$, with $r_1 \neq R_1 := \sigma \|\beta_1\|$, such that $\mathcal{T}u \nleq u$ for all $u \in K$
        with $\|u\| = r_1$.

Then the map $\mathcal{T}$ has, at least, a non-zero fixed point $u_1 \in K \backslash \{u \in K; \|u\| = r_1\}$
such that

$$\min\{r_1, R_1\} \leq \|u_1\| \leq \max\{r_1, R_1\}.$$

Proof. By the property (4.16) we have that if $u \in K$ with $\|u\| = R_1$ then
$u \geq \beta_1$. The definition of R_1 says us that u, $\beta_1 \in K_4$, so, since $\mathcal{T}$ is monotone
non-increasing on K_4 we have

$$\mathcal{T}u \leq \mathcal{T}\beta_1 \leq \beta_1 \leq u \text{ for all } u \in K \text{ with } \|u\| = R_1.$$

If $\mathcal{T}u = u$, we have a fixed point such that $\|u\| = R_1$, on the contrary,
$\mathcal{T}u \nleq u$ for all $u \in K$ such that $\|u\| = R_1$ which, together with (T8), implies
by Theorems 40 and 41, the existence of a fixed point with the desired
localization property. □

Remark 41. (i) Note that if $\mathcal{T}\beta_1 \neq \beta_1$, in the proof of the two previous
results it is showed that the map $\mathcal{T}$ has, at least, a non-zero fixed point
$u_1 \in K$, such that

$$\min\{r_1, R_1\} < \|u_1\| < \max\{r_1, R_1\}.$$

The same conclusion holds if in condition (4.16), we assume that the
constant $\sigma > 1$. To verify this, it is enough to take into account that if there
is a fixed point $u \in K$ with $\|u\| = R_1$ then $\mathcal{T}u \leq \mathcal{T}\beta_1 \leq \beta_1 \leq u$ and, as a
consequence, $\sigma \|\beta_1\| = \|u\| = \|\beta_1\|$.

(ii) An example of cone satisfying condition (4.16) is for instance the one
used in [9] (with $f(t) = t^2$ on $[0, 1]$)

$$K = \{cf(t) : f \in L^\infty(I), f \geq 0 \text{ a.e. } t \in I, \|f\|_1 > 0, c \geq 0\}.$$

Lemma 48. *Condition (4.16) is equivalent to the following one:*

there exists $\sigma \geq 1$ such that for $x, y \in K$ with $\|x\| \geq \sigma \|y\|$ we have $x \geq y$.

(4.17)

Proof. Obviously if condition (4.17) is fulfilled then (4.16) also holds.

Suppose now that (4.16) is satisfied, and let x, $y \in K$ be such that $\|x\| \geq \sigma \|y\|$, in consequence there is $\lambda \geq 1$ such that $\|x/\lambda\| = \sigma \|y\|$. Condition (4.16) shows that $x \geq \lambda y \geq y$, i.e., condition (4.17) holds. This proves the result. $\square$

In an analogous way to Corollary 43, we arrive at the following "dual result".

Corollary 44. *Let X be a real Banach space, K a cone satisfying condition (4.16) and $\mathcal{T}: K \to K$ a completely continuous operator. Assume that*
(T9)   there exists $\alpha_1 \in K$, $\alpha_1 \neq 0$ such that $\alpha_1 \leq \mathcal{T}\alpha_1$,
(T10)   the map $\mathcal{T}$ is monotone non-increasing in the set

$$K_5 = \{u \in K : \min\{\|\alpha_1\|, \|\mathcal{T}\alpha_1\|\} \leq \|u\| \leq \sigma \|\mathcal{T}\alpha_1\|\},$$

(T11)   there exists $r_1 > 0$, with $r_1 \neq R_1 := \sigma \|\mathcal{T}\alpha_1\|$, such that $\mathcal{T}u \not\leq u$ for all $u \in K$ with $\|u\| = r_1$.
Then the map $\mathcal{T}$ has, at least, a non-zero fixed point $u_1 \in K\backslash\{u \in K; \|u\| = r_1\}$ such that

$$\min\{r_1, R_1\} \leq \|u_1\| \leq \max\{r_1, R_1\}.$$

Proof. By the property (4.16) we have that if $u \in K$ with $\|u\| = R_1$ then $u \geq \mathcal{T}\alpha_1$.

Suppose now that $\|\alpha_1\| \geq \sigma \|\mathcal{T}\alpha_1\|$. From Lemma 48, we have that $\alpha_1 \geq \mathcal{T}(\alpha_1)$, which implies that α_1 is a fixed point with $\|\alpha_1\| = R_1$ and $\sigma = 1$, so the result holds.

When $\|\alpha_1\| < \sigma \|\mathcal{T}\alpha_1\|$, from the definition of R_1 it is obvious that α_1, $\mathcal{T}(\alpha_1) \in K_5$, so, since $\mathcal{T}$ is monotone non-increasing on K_5 we have

$$\mathcal{T}(\mathcal{T}(\alpha_1)) \geq \mathcal{T}(\alpha_1) > 0,$$

and the result holds from Theorem 59. $\square$

Remark 42. Note that if $\mathcal{T}(\alpha_1) \neq \alpha_1$ or $\sigma > 1$, in the proof of the previous result it is showed that the map $\mathcal{T}$ has, at least, a non-zero fixed point

$u_1 \in K$, such that

$$\min\{r_1, R_1\} < \|u_1\| < \max\{r_1, R_1\}.$$

4.2.6 Non-decreasing Operators

For monotone non-decreasing operators, in case of an upper solution, several fixed point theorems for normal and solid cones have been proved in Subsection 4.2.4 (see Theorems 53, 54, and 55). In the sequel, we prove a fixed point result in this direction for a nonnecessarily normal cone that satisfies (4.16).

Now, we give a result under the assumption of the existence of a lower solution.

Theorem 60. *Let X be a real Banach space, K a normal cone (not necessarily solid) with normal constant $c \geq 1$, that satisfies condition (4.16), and $\mathcal{T}: K \to K$ a completely continuous operator. Assume that there is a lower solution as in (T9), and*

(T12)   the map $\mathcal{T}$ is monotone non-decreasing in the set

$$K_6 = \left\{ u \in K : \frac{\|\alpha_1\|}{c} \leq \|u\| \leq c\sigma \|\alpha_1\| \right\},$$

(T13)   there exists $r_1 > 0$, with $r_1 \neq R_1 := \sigma \|\alpha_1\|$, such that $\mathcal{T}u \not\geq u$ for all $u \in K$ with $\|u\| = r_1$.
Then the map $\mathcal{T}$ has, at least, a non-zero fixed point u_1 in K, that either belongs to K_6 or is such that

$$\min\{r_1, R_1\} < \|u_1\| < \max\{r_1, R_1\}.$$

Proof. By the property (4.16) we have that if $u \in K$ with $\|u\| = R_1$ then $u \geq \alpha_1$.

Suppose first that we can choose $\beta_1 \in K$ with $\|\beta_1\| = R_1$ and $\mathcal{T}\beta_1 \leq \beta_1$. Since $\alpha_1 \leq \beta_1$ and due to the normality of the cone K we have that $[\alpha_1, \beta_1] \subset K_6$, which implies that $\mathcal{T}$ is monotone non-decreasing on $[\alpha_1, \beta_1]$. Then we can apply Theorem 51 to ensure the existence of the extremal fixed points of $\mathcal{T}$ on $[\alpha_1, \beta_1]$.

Now suppose that $\mathcal{T}u \not\leq u$ for all $u \in K$ with $\|u\| = R_1$. By (T13) there exists $r_1 > 0$ such that $\mathcal{T}u \not\geq u$ for all $u \in K$ with $\|u\| = r_1$. Therefore by Theorems 40 and 41, there exists a non-zero fixed point u_1 in the required set. □

Remark 43. We stress that the above Theorem can be combined to prove the existence of multiple fixed points. The idea is to use a nesting argument similar to those utilized for example in [51,61], where the authors used the index theory, and in [43,53], where Theorems 40 and 41 were used. In Subsection 4.2.6.1 we do this in the case of the existence of two non-trivial fixed points and we refer to Theorem 3.4 of [39] to give an idea of the type of results that may be stated in the case of n fixed points.

4.2.6.1 Multiplicity of Solutions

In this subsection we apply the results obtained in Subsections 4.2.5 and 4.2.6 to deduce the existence of at least two non-trivial solutions of a periodic boundary value problem. The arguments follow the ideas given in [13, Section 4] for a fourth order problem that models the displacement of a beam with feedback controllers.

So, we are interested in ensuring the existence of positive solutions of the periodic Hill's equation

$$u''(t)+a(t)\,u(t)=\lambda g(t)f(u(t)),\ t\in I,\quad u(0)=u(T),\ u'(0)=u'(T),\quad (4.18)$$

where $g\in L^1(I)$, $g\geq 0$ a.e. I and $f:[0,\infty)\to[0,\infty)$ is a continuous function.

Here the potential a satisfies condition $(a0)$, i.e., from Lemma 24 and Theorem 15, it must satisfy that

$$\lambda_0^P[a] < 0 < \lambda_0^A[a]. \qquad (4.19)$$

As we have noticed, the solutions of the studied problem coincide with the fixed points of the integral equation

$$u(t) = \lambda \int_0^T G_P[a](t,s)\,g(s)f(u(s))\,ds := \mathcal{T}u(t), \quad t\in I.$$

As usual, we denote

$$0 < m := \min_{t,s\in I} G_P[a](t,s) < \max_{t,s\in I} G_P[a](t,s) =: M$$

and

$$0 < \gamma \leq \frac{m}{M}$$

to be fixed.

As in the previous subsection, it is routine to prove that $\mathcal{T}: C(I) \to C(I)$ leaves invariant the cone

$$K = \left\{ u \in C(I) : \min_{t \in I} u(t) \geq \gamma \|u\|_\infty \right\},$$

which is a normal solid cone with normal constant $c = 1$.

We will make use of the numbers

$$\delta_* = \min_{t \in I} \int_0^T G_P[a](t,s)g(s)\,ds, \quad \delta^* = \max_{t \in I} \int_0^T G_P[a](t,s)g(s)\,ds.$$

Now we present the main result of this subsection.

Theorem 61. *Assume that (4.19) holds. Let g be such that $\delta_* > 0$, and choose $0 < \gamma \leq m/M$ satisfying $\gamma \delta^* < \delta_*$.*

Moreover let the real constants $\beta_1, \alpha_2, R_1, R_2 > 0$ be such that

$$\alpha_2 \geq \frac{1+\gamma}{1-\gamma} R_2, \quad \beta_1 \geq \frac{1+\gamma}{1-\gamma} R_1, \quad and \ \gamma R_2 > R_1.$$

Moreover, we assume the following conditions
(i) f is monotone non-decreasing on $[\gamma R_1, \beta_1]$,
(ii) f is monotone non-increasing on $[\gamma R_2, \alpha_2]$,
(iii) $\displaystyle\lim_{u \to 0^+} \frac{f(u)}{u} = +\infty$ and $\displaystyle\lim_{u \to +\infty} \frac{f(u)}{u} = 0$.
Then the periodic boundary value problem (4.18) has at least two positive solutions for any $\lambda > 0$ satisfying

$$\frac{\alpha_2(1-\gamma)}{f(\alpha_2)(\delta_* - \gamma \delta^*)} \leq \lambda \leq \frac{\beta_1(1-\gamma)}{f(\beta_1)(\delta^* - \gamma \delta_*)}.$$

Proof. The main idea in the proof is to apply Theorems 58 and 55 in two disjoint conical shells in order to get two different non-trivial fixed points.

Firstly we are going to check that conditions of Theorem 58 are satisfied. To this end we denote "$\preceq$" as the order induced in $C(I)$ by the cone K, i.e.,

$$x \preceq y \iff \min_{t \in I}(y(t) - x(t)) \geq \gamma \|y - x\|_\infty.$$

(1.a) $\alpha_2 \preceq \mathcal{T}\alpha_2$ and there exists $R_2 > 0$ such that $B[\alpha_2, R_2] \subset K$.
    We have $\alpha_2 \preceq \mathcal{T}\alpha_2$ because of the inequality

$$\frac{\alpha_2(1-\gamma)}{f(\alpha_2)(\delta_* - \gamma \delta^*)} \leq \lambda.$$

On the other hand, the inequality $\alpha_2 \geq \frac{1+\gamma}{1-\gamma} R_2$ implies that $B[\alpha_2, R_2] \subset K$. Indeed, let $u \in B[\alpha_2, R_2]$, that is,

$$\alpha_2 - R_2 \leq u(t) \leq \alpha_2 + R_2 \quad \text{for all } t \in [0, 1].$$

Since $\alpha_2 > R_2$ we have that $u(t) > 0$ for all $t \in [0, 1]$. Moreover it is easy to check that

$$\min_{t \in [0,1]} u(t) \geq \alpha_2 - R_2 \geq \gamma(\alpha_2 + R_2) \geq \gamma \|u\|,$$

which means that $u \in K$.

(1.b)   *The map $\mathcal{T}$ is monotone non-increasing (with the partial order $\preceq$) in the set*

$$K_2' = \{u \in K : R_2 \leq \|u\| \leq \alpha_2\}.$$

This fact is consequence of the assumption (*ii*).

(1.c)   *There exists $r_2 > 0$, with $r_2 \neq R_2$, such that $\mathcal{T}u \npreceq u$ for all $u \in K$ with $\|u\| = r_2$.*
By the second part of the assumption (*iii*) we can ensure this assertion for $r_2 > R_2$ large enough.

So Theorem 58 implies the existence of a solution $u_2 \in K$ such that

$$R_2 \leq \|u_2\| < r_2.$$

Now, we are going to check that conditions of Theorem 55 are satisfied:

(2.a) $\mathcal{T}\beta_1 \preceq \beta_1$ *and there exists $R_1 > 0$ such that $B[\beta_1, R_1] \subset K$.*
We have $\mathcal{T}\beta_1 \preceq \beta_1$ because of the inequality

$$\lambda \leq \frac{\beta_1 \, (1-\gamma)}{f(\beta_1) \, (\delta^* - \gamma \, \delta_*)}.$$

Again $\beta_1 \geq \frac{1+\gamma}{1-\gamma} R_1$ implies that $B[\beta_1, R_1] \subset K$ by reasoning as in the proof of claim (1.a).

(2.b)   *The map $\mathcal{T}$ is monotone non-decreasing (with the partial order $\preceq$) in the set*

$$K_1' = \{u \in K : R_1 \leq \|u\| \leq \beta_1\}.$$

This fact is consequence of the assumption (i).

(2.c)   *There exists $r_1 > 0$, with $r_1 \neq R_1$, such that $\mathcal{T}u \npreceq u$ for all $u \in K$ with $\|u\| = r_1$.*
By the first part of the assumption (*iii*) we have this result for $0 < r_1 < R_1$ small enough.

Then applying Theorem 55 we get the existence of a solution $u_1 \in K$ such that $r_1 < \|u_1\| \le R_1$. Since $R_1 < \gamma R_2$ we have that $u_1 \ne u_2$ and the result is proven. $\square$

The following example illustrates our previous theorem.

Example 12. Consider now problem (4.18), with $a(t) \ge 0$ a.e. $t \in I$ satisfying (4.19), g such that $c_1 a(t) \le g(t) \le c_2 a(t)$ a.e. $t \in I$ for some constants $0 < c_1 \le c_2$, and

$$f(x) = \sqrt{x} + \frac{300e^{-e^{20(2-x)}}\left(\tan^{-1}(21-x) + \frac{\pi}{2}\right)}{\pi}.$$

Graph of f on $[0,1]$ Graph of f on $[1,3]$ Graph of f on $[3,30]$

In this case, as it is proved in [8, Proposition 1.4.10], we know that

$$\int_0^T G_P[a](t,s)\, a(s)\, ds = 1$$

so, as a consequence, we have that $c_1 \le \delta_* \le \delta^* \le c_2$.

Now, it is easy to check that, choosing $\gamma < c_1/c_2$ and, if it satisfies

$$\frac{1}{4} \le \gamma \le \frac{3}{5} \tag{4.20}$$

all the assumptions of Theorem 61 are satisfied by taking $\beta_1 = 3/2$, $\alpha_2 = 20$, $R_1 = 1/4$, and $R_2 = 20\frac{1-\gamma}{1+\gamma}$.

So, by Theorem 61, our considered problem (4.18), has at least two positive solutions, u_1 and u_2, provided that

$$0.0871566\,\frac{1-\gamma}{c_1 - \gamma c_2} \le \lambda \le 1.22474\,\frac{1-\gamma}{c_2 - \gamma c_1}$$

and they satisfy

$$\|u_1\| \le 1/4 < 20\,\frac{1-\gamma}{1+\gamma} \le \|u_2\|.$$

In the particular case of $a(t) = k^2$, as we have noticed in Chapter 3, condition (4.19) is rewritten as $0 < k < \pi/T$. In such a case, using the expression showed in Example 6, it is immediate to verify that

$$\gamma = \cos\left(\frac{kT}{2}\right),$$

and inequality (4.20) reads as follows:

$$1.85459 \leq k\,T \leq 2.63623.$$

4.2.7 Problems with Parametric Dependence

In this subsection we obtain existence, nonexistence, and multiplicity results for the Hill's equation coupled with periodic boundary conditions. The proof studies the cases where the related Green's function is non-negative, so, the results showed in Chapter 3 are fundamental to treat this situation. The approach strongly depends on the construction of suitable cones where it is possible to find the fixed points of the constructed integral operator. In particular, the used one is a variant of the Krasnosel'skiĭ's fixed point Theorem (Theorem 42) presented in Subsection 4.2.2. Such variant allows us to consider nonlinear parts with singularities. The results exposed here are proved in [11].

As we have noted in previous subsection, the obtained results when the Green's function is strictly positive on $I \times I$ can be directly adapted for Neumann conditions. The case for strictly negative Green's function can be also treated for different boundary conditions, as the ones considered in Chapter 3, it is enough to multiply times -1 both sides of the equation, and we have that the Green's function remains strictly positive. However, contrary to some situations in previous subsection, as we will see, the results obtained in Subsection 4.2.7.2 are not applicable to Dirichlet or mixed problems since, in such a case (with the notation used there), condition $\widetilde{(H1)}$ does not hold because $\beta = 0$.

4.2.7.1 Introduction and Preliminaries

In [41] the authors obtained existence, multiplicity, and nonexistence results for the periodic problem

$$\begin{aligned}
u''(t) - k^2 u(t) &= \lambda\, g(t) f(u(t)), \quad t \in [0, 2\pi] \\
u(0) &= u(2\pi), \quad u'(0) = u'(2\pi),
\end{aligned} \qquad (4.21)$$

depending on a real parameter $\lambda > 0$. Although it is not explicitly mentioned in [41] the main arguments are valid because the related Green's function of (4.21) is strictly negative for all $k > 0$.

This subsection is devoted to give complementary results to those of [41] for the case of a non-negative related Green's function. In particular we shall deal with problem

$$u''(t) + a(t) u(t) = \lambda g(t) f(u(t)) + c(t), \quad \text{a.e. } t \in I$$
$$u(0) = u(T), \quad u'(0) = u'(T), \tag{4.22}$$

assuming that its Green's function is non-negative (for instance, as we have seen in Chapter 3, if $a(t) = k^2$ this means $0 < k \le \frac{\pi}{T}$). Moreover, in order to give wider applicable results we will also allow $f(x)$ to be singular at $x = 0$ (the reader may have in mind the model $f(x) = \dfrac{1}{x^\alpha}$, for some $\alpha > 0$).

The main tool used will be the Krasnosel'skiĭ's fixed point theorem in a cone, which has been introduced in Subsection 4.2.2 and that follows as a consequence of the degree theory.

Along the subsection, we will use cones of the form

$$K = \{u \in C(I), \ u \ge 0 \text{ on } I : \ \varphi(u) \ge \sigma \|u\|\},$$

where $0 < \sigma \le 1$ is a fixed constant and $\varphi \ge 0$ is a functional defined on $X = \{u \in C(I), \ u \ge 0 \text{ on } I\}$ that satisfies
- $\varphi(x + y) \ge \varphi(x) + \varphi(y)$ for all $x, y \in X$,
- $\varphi(\lambda x) = \lambda \varphi(x)$ for all $\lambda > 0$ and $x \in X$.

In particular, in Subsection 4.2.7.2 we will use the standard choice $\varphi(x) = \min_{t \in I} u(t)$ and in Subsection 4.2.7.3 we will use $\varphi(u) = \int_0^T u(s)\,ds$, which has been introduced in [42].

The potential a is chosen so that the linear problem (4.2) is nonresonant.

So, we can ensure the existence of $G_P[a]$, its related Green's function, and define, for each $\lambda > 0$ the operator

$$\mathcal{T}_\lambda : \mathcal{D}(\mathcal{T}_\lambda) \equiv \{u \in C(I) : u(t) > 0 \text{ for all } t \in I\} \to C(I)$$

given by

$$\mathcal{T}_\lambda u(t) = \lambda \int_0^T G_P[a](t, s) g(s) f(u(s))\,ds + \int_0^T G_P[a](t, s)\, c(s)\,ds, \quad t \in I.$$

$$\tag{4.23}$$

As a consequence, we have that $u > 0$ is a solution of problem (4.22) if and only if $u = T_\lambda u$.

Throughout this subsection we shall use the following notation:

$$\gamma(t) = \int_0^T G_P[a](t, s)\, c(s)\, ds,$$

$$m = \min_{t,s \in I} G_P[a](t, s) \quad \text{and} \quad M = \max_{t,s \in I} G_P[a](t, s),$$

and

$$\overline{f_0} = \lim_{x \to 0^+} \frac{f(x)}{x} \quad \text{and} \quad \overline{f_\infty} = \lim_{x \to \infty} \frac{f(x)}{x}.$$

Moreover, for an essentially bounded function $h : I \to \mathbb{R}$ we define

$$h_* = \inf_{t \in I} \operatorname{ess} h(t) \quad \text{and} \quad h^* = \sup_{t \in I} \operatorname{ess} h(t).$$

4.2.7.2 Positive Green's Function

As a first approach to this problem, we assume the following hypotheses:

(H0) $\gamma_* > 0$ or $c(t) \equiv 0$.

(H1) $a \in L^p(I)$, $p \geq 1$, is such that condition (a0) is fulfilled.

(H2) $g \in L^1(I)$, $g(t) \geq 0$ a.e. $t \in I$ and $\int_0^T g(s)\, ds > 0$.

(H3) $f : (0, \infty) \to (0, \infty)$ is continuous.

(H4) $c \in L^1(I)$.

Notice that condition (H3) allows f to be singular at $x = 0$. In particular (H3) is satisfied when $f(x) = \frac{1}{x^\alpha}$, $\alpha > 0$ (the case $0 < \alpha < 1$ is called a weak singularity while $\alpha \geq 1$ is called strong singularity).

From (H1) it follows that $m > 0$ and we define the cone

$$K := \left\{ u \in C(I) : \min_{t \in I} u(t) \geq \sigma \|u\|_\infty \right\},$$

where

$$\sigma = \min \left\{ \frac{m}{M}, \frac{\gamma_*}{\gamma^*} \right\} \quad \text{if } \gamma_* > 0 \quad \text{or} \quad \sigma = \frac{m}{M} \quad \text{if } c(t) \equiv 0.$$

In both cases $0 < \sigma < 1$ and for $0 < r < R$ we define

$$K_{r,R} := \{ u \in K : r \leq \|u\|_\infty \leq R \}.$$

Next we give sufficient conditions for the solvability of problem (4.22).

Theorem 62. *Assume that conditions* (H0), (H1), (H2), (H3), *and* (H4) *are fulfilled. Then for each* $\lambda > 0$ *and* $0 < r < R$ *the operator* $T_\lambda : K_{r,R} \to K$ *given by (4.23) is well defined and completely continuous.*

Moreover, if either

(i) $\|T_\lambda u\|_\infty \leq \|u\|_\infty$ *for any* $u \in K$ *with* $\|u\|_\infty = r$ *and* $\|T_\lambda u\|_\infty \geq \|u\|_\infty$ *for any* $u \in K$ *with* $\|u\|_\infty = R$, *or*

(ii) $\|T_\lambda u\|_\infty \geq \|u\|_\infty$ *for any* $u \in K$ *with* $\|u\|_\infty = r$ *and* $\|T_\lambda u\|_\infty \leq \|u\|_\infty$ *for any* $u \in K$ *with* $\|u\|_\infty = R$,

then T_λ *has a fixed point in* $K_{r,R}$ *which is a positive solution of problem (4.22).*

Proof. Note that if $u \in K_{r,R}$ then $0 < \sigma r \leq u(t) \leq R$ for all $t \in I$, so $K_{r,R} \subset \mathcal{D}(T_\lambda)$ and then $T_\lambda : K_{r,R} \to \mathcal{C}(I)$ is well defined. Following the arguments showed in previous subsections, we can verify that $T_\lambda(\mathcal{D}(T_\lambda)) \subset K$ and that T_λ is completely continuous. Then, from Krasnosel'skiĭ's fixed point Theorem (Theorem 42) it follows the existence of a fixed point of T_λ in $K_{r,R}$ which is, by the definition of T_λ, a positive solution of problem (4.22). $\square$

Before proving the existence and multiplicity results for problem (4.22) we need to use some technical lemmas that will be proved in the next subsection.

Auxiliary Results

Lemma 49. *Assume that conditions* (H0), (H1), (H2), (H3), *and* (H4) *are satisfied. Then, for each* $R > \gamma^*$ *there exists* $\lambda_0(R) > 0$ *such that for every* $0 < \lambda \leq \lambda_0(R)$ *we have*

$$\|T_\lambda u\|_\infty \leq \|u\|_\infty \quad \text{for } u \in K \text{ with } \|u\|_\infty = R.$$

Proof. Fix $R > \gamma^*$ and let $u \in K$ with $\|u\|_\infty = R$. If

$$0 < \lambda \leq \lambda_0(R) := \frac{R - \gamma^*}{M \max_{u \in [\sigma R, R]} f(u) \int_0^T g(s)\, ds}$$

then, for all $t \in I$ the following inequalities hold:

$$T_\lambda u(t) = \lambda \int_0^T G_P[a](t, s) g(s) f(u(s))\, ds + \gamma(t)$$

$$\leq \lambda M \max_{u \in [\sigma R, R]} f(u) \int_0^T g(s)\, ds + \gamma^* \leq R = \|u\|_\infty,$$

and thus $\|T_\lambda u\|_\infty \leq \|u\|_\infty$. $\square$

Lemma 50. *Assume that conditions* $(H0), (H1), (H2), (H3),$ *and* $(H4)$ *are fulfilled. Then, for each* $r > 0$ *there exists* $\lambda_0(r) > 0$ *such that for every* $\lambda \geq \lambda_0(r)$ *we have*

$$\|T_\lambda u\|_\infty \geq \|u\|_\infty \quad \textit{for } u \in K \textit{ with } \|u\|_\infty = r.$$

Proof. Fix $r > 0$ and let $u \in K$ with $\|u\|_\infty = r$. If

$$\lambda \geq \lambda_0(r) := \frac{r}{m \, \min_{u \in [\sigma r, r]} f(u) \int_0^T g(s) \, ds}$$

then

$$
\begin{aligned}
T_\lambda u(t) &= \lambda \int_0^T G_P[a](t, s) g(s) f(u(s)) \, ds + \gamma(t) \\
&\geq \lambda \, m \, \min_{u \in [\sigma R, R]} f(u) \int_0^T g(s) \, ds + \gamma_* \geq r = \|u\|_\infty,
\end{aligned}
$$

and thus $\|T_\lambda u\|_\infty \geq \|u\|_\infty$. $\quad\quad\square$

Lemma 51. *Suppose that conditions* $(H1), (H2), (H3), (H4)$ *are satisfied and* $c(t) \equiv 0$. *Then, if* $\overline{f_0} = 0$ *there exists* $r_0(\lambda) > 0$ *such that for every* $0 < r \leq r_0(\lambda)$ *we have*

$$\|T_\lambda u\|_\infty \leq \|u\|_\infty \quad \textit{for } u \in K \textit{ with } \|u\|_\infty = r.$$

Proof. Since $\overline{f_0} = 0$ for $\varepsilon = \varepsilon(\lambda) = \frac{1}{\lambda M \int_0^T g(s) \, ds}$, there exists $r_0(\lambda) > 0$ such that $f(u) \leq \varepsilon u$ for each $0 < u \leq r_0(\lambda)$.

Fix $0 < r \leq r_0(\lambda)$ and let $u \in K$ with $\|u\|_\infty = r$. Then

$$
\begin{aligned}
T_\lambda u(t) &= \lambda \int_0^T G_P[a](t, s) g(s) f(u(s)) \, ds \leq \lambda M \int_0^T g(s) \varepsilon u(s) \, ds \\
&\leq \lambda M \varepsilon \|u\|_\infty \int_0^T g(s) \, ds = \|u\|_\infty,
\end{aligned}
$$

and thus $\|T_\lambda u\|_\infty \leq \|u\|_\infty$. $\quad\quad\square$

Lemma 52. *Assume that hypotheses* $(H0), (H1), (H2), (H3),$ *and* $(H4)$ *hold. Then, if* $\overline{f_0} = \infty$ *there exists* $r_0(\lambda) > 0$ *such that for every* $0 < r \leq r_0(\lambda)$ *we have*

$$\|T_\lambda u\|_\infty \geq \|u\|_\infty \quad \textit{for } u \in K \textit{ with } \|u\|_\infty = r.$$

Proof. Since $\overline{f_0} = \infty$ for $L = L(\lambda) = \frac{1}{\lambda\, m\, \sigma\, \int_0^T g(s)\, ds}$ there exists $r_0(\lambda) > 0$ such that $f(u) \geq Lu$ for each $0 < u \leq r_0(\lambda)$.

Fix $0 < r \leq r_0(\lambda)$ and let $u \in K$ with $\|u\|_\infty = r$. Then

$$
\begin{aligned}
\mathcal{T}_\lambda u(t) &= \lambda \int_0^T G_P[a](t,s)\, g(s) f(u(s))\, ds + \gamma(t) \geq \lambda m \int_0^T g(s)\, Lu(s)\, ds + \gamma_* \\
&\geq \lambda m L \sigma \|u\|_\infty \int_0^T g(s)\, ds = \|u\|_\infty,
\end{aligned}
$$

and thus $\|\mathcal{T}_\lambda u\|_\infty \geq \|u\|_\infty$. $\qquad\square$

Lemma 53. *Suppose that conditions $(H0)$, $(H1)$, $(H2)$, $(H3)$, and $(H4)$ are satisfied. Then, if $\overline{f_\infty} = 0$ then there exists $R_0(\lambda) > 0$ such that for every $R \geq R_0(\lambda)$ we have*

$$\|\mathcal{T}_\lambda u\|_\infty \leq \|u\|_\infty \quad \text{for } u \in K \text{ with } \|u\|_\infty = R.$$

Proof. Since $\overline{f_\infty} = 0$ for $\varepsilon(\lambda) = \frac{1}{2\lambda\, M\, \int_0^T g(s)\, ds}$ there exists $R_1(\lambda) > 0$ such that $f(u) \leq \varepsilon u$ for each $u \geq R_1(\lambda)$. We define $R_0(\lambda) := \max\{R_1(\lambda)/\sigma, 2\gamma^*\}$.

Fix $R \geq R_0(\lambda)$ and let $u \in K$ with $\|u\|_\infty = R$. Then

$$
\begin{aligned}
\mathcal{T}_\lambda u(t) &= \lambda \int_0^T G_P[a](t,s)\, g(s) f(u(s))\, ds + \gamma(t) \leq \lambda M \int_0^T g(s)\varepsilon u(s) + \gamma^* \\
&\leq \lambda M \varepsilon \|u\|_\infty \int_0^T g(s)\, ds + \gamma^* = \frac{R}{2} + \gamma^* \leq \frac{R}{2} + \frac{R}{2} = R = \|u\|_\infty,
\end{aligned}
$$

and thus $\|\mathcal{T}_\lambda u\|_\infty \leq \|u\|_\infty$. $\qquad\square$

Lemma 54. *Assume that $(H0)$, $(H1)$, $(H2)$, $(H3)$, and $(H4)$ are fulfilled. Then, if $\overline{f_\infty} = \infty$ there exists $R_0(\lambda) > 0$ such that for every $R \geq R_0(\lambda)$ we have*

$$\|\mathcal{T}_\lambda u\|_\infty \geq \|u\|_\infty \quad \text{for } u \in K \text{ with } \|u\|_\infty = R.$$

Proof. Since $\overline{f_\infty} = \infty$ for $L = L(\lambda) = \frac{1}{\lambda\, m\, \sigma\, \int_0^T g(s)\, ds}$ there exists $R_1(\lambda) > 0$ such that $f(u) \geq Lu$ for each $u \geq R_1(\lambda)$. We define $R_0(\lambda) := R_1(\lambda)/\sigma$.

Fix $R \geq R_0(\lambda)$ and let $u \in K$ with $\|u\|_\infty = R$. Then

$$
\begin{aligned}
\mathcal{T}_\lambda u(t) &= \lambda \int_0^T G_P[a](t,s)\, g(s) f(u(s))\, ds + \gamma(t) \geq \lambda m \int_0^T g(s)\, Lu(s)\, ds + \gamma_* \\
&\geq \lambda m L \sigma \|u\|_\infty \int_0^T g(s)\, ds = \|u\|_\infty,
\end{aligned}
$$

and thus $\|\mathcal{T}_\lambda u\|_\infty \geq \|u\|_\infty$. $\qquad\square$

In the sequel, we study separately the two different cases considered in condition $(H0)$, i.e., $\gamma_* > 0$ or $c(t) \equiv 0$.

The case $\gamma_* > 0$

Theorem 63. *Assume that conditions* $(H1), (H2), (H3),$ *and* $(H4)$ *are fulfilled. If moreover* $\gamma_* > 0$ *the following results hold:*

1. *There exists* $\lambda_0 > 0$ *such that problem (4.22) has a positive solution if* $0 < \lambda < \lambda_0$.
2. *If* $\overline{f_\infty} = 0$ *then problem (4.22) has a positive solution for every* $\lambda > 0$.
3. *If* $\underline{f_\infty} = \infty$ *then there exists* $\lambda_0 > 0$ *such that problem (4.22) has two positive solutions if* $0 < \lambda < \lambda_0$.
4. *If* $\underline{f_0} > 0$ *and* $\underline{f_\infty} > 0$ *then there exists* $\lambda_0 > 0$ *such that problem (4.22) has no positive solutions if* $\lambda > \lambda_0$.

Proof. Fix $0 < r < \gamma_*$. Then for each $\lambda > 0$ and $u \in K$ with $\|u\|_\infty = r$ we have

$$\|T_\lambda u\|_\infty \geq T_\lambda u(t) = \lambda \int_0^T G_P[a](t, s) g(s) f(u(s)) \, ds + \gamma(t) \geq \gamma_* > r = \|u\|_\infty.$$

Part 1. Fix $R > \gamma^* (\geq \gamma_* > r)$ and take $\lambda_0 = \lambda_0(R)$ given by Lemma 49. Then from Theorem 42 (ii) it follows the existence of a positive solution for problem (4.22) if $0 < \lambda < \lambda_0$.

Part 2. Fix $\lambda > 0$ and take $R > \max\{r, R_0(\lambda)\}$, where $R_0(\lambda)$ is given by Lemma 53. Then from Theorem 42 (ii) it follows the existence of a positive solution for problem (4.22).

Part 3. Fix $R_2 > R_1 > \gamma^* (\geq \gamma_* > r)$ and take $\lambda_0 = \min\{\lambda_0(R_1), \lambda_0(R_2)\}$, where $\lambda_0(R_1)$ and $\lambda_0(R_2)$ are given by Lemma 49.

Now, fix $0 < \lambda < \lambda_0$ and take $R > \max\{R_2, R_0(\lambda)\}$, where $R_0(\lambda)$ is given by Lemma 54. Therefore from Theorem 42 it follows the existence of two positive solutions u_1 and u_2 for problem (4.22) such that

$$r \leq \|u_1\| \leq R_1 < R_2 \leq \|u_2\| \leq R.$$

Part 4. Since $\underline{f_0} > 0$ and $\underline{f_\infty} > 0$ there exists $L > 0$ such that $f(u) \geq L u$ for all $u > 0$. Define

$$\lambda_0 := \frac{1}{m \sigma L \int_0^T g(s) \, ds}.$$

If for $\lambda > \lambda_0$ there exists a positive solution u of problem (4.22) we know that $u \in \mathcal{D}(\mathcal{T}_\lambda)$ and, as a consequence, $u = \mathcal{T}_\lambda(u) \in K$. Therefore we deduce the following inequalities

$$\|u\|_\infty = \|\mathcal{T}_\lambda u\|_\infty \geq \mathcal{T}_\lambda u(t) \;=\; \lambda \int_0^T G_P[a](t, s)\, g(s)\, f(u(s))\, ds + \gamma(t)$$

$$\geq \;\lambda\, m \int_0^T g(s)\, L u(s)\, ds + \gamma_*$$

$$\geq \;\lambda\, m\, L\, \sigma \|u\|_\infty \int_0^T g(s)\, ds$$

$$> \;\|u\|_\infty,$$

and we attain a contradiction. □

Example 13. Let us consider the forced Mathieu-Duffing type equation

$$u'' + (e + b\cos(t))\, u - \lambda\, u^3 = c(t), \tag{4.24}$$

with $e \geq 0$ and $b \in \mathbb{R}$.

This equation fits into expression (4.22) by defining $a(t) = e(1 + b\cos(t))$, $g(t) = 1$ and $f(x) = x^3$.

Eq. (4.24), with $c(t) \equiv 0$, was studied in [37] where a sufficient condition for the existence of a 2π-periodic solution is given. However, since the proof relies on the application of Schauder's fixed point theorem in a ball centered at the origin, the trivial solution $u(t) \equiv 0$ is not excluded. The existence of a non-trivial solution was later obtained by Torres in [94, Corollary 4.2]. More precisely, Torres proved that if function $a(t) > 0$ a.e. $t \in [0, 2\pi]$ and $\|a\|_p < K(2p^*, 2\pi)$ ($K(\alpha, T)$ introduced in (3.3) and obtained in Appendix A), then the homogeneous problem (with $c(t) \equiv 0$ in (4.24)) has at least two non-trivial one-signed 2π-periodic solutions.

As a consequence of Example 13 and Theorem 63, Part 3, we arrive at the following multiplicity result for the nonhomogeneous ($c(t) \not\equiv 0$) equation (4.24) with a not necessarily constant sign function a.

Corollary 45. *If condition (3.19) is satisfied and $\gamma_* > 0$ then there exists $\lambda_0 > 0$ such that Eq. (4.24) has at least two positive 2π-periodic solutions provided that $0 < \lambda < \lambda_0$.*

The case $c(t) \equiv 0$

Theorem 64. *Assume that conditions* (H1), (H2), (H3), *and* (H4) *hold. If moreover* $c(t) \equiv 0$ *the following results hold:*

1. *If* $\overline{f_0} = \infty$ *or* $\overline{f_\infty} = \infty$ *then there exists* $\lambda_0 > 0$ *such that problem (4.22) has a positive solution if* $0 < \lambda < \lambda_0$.
2. *If* $\overline{f_\infty} = 0$ *then there exists* $\lambda_0 > 0$ *such that problem (4.22) has a positive solution for every* $\lambda > \lambda_0$.
3. *If* $\overline{f_0} = \infty$ *and* $\overline{f_\infty} = 0$ *then problem (4.22) has a positive solution for every* $\lambda > 0$.
4. *If* $\overline{f_0} = \infty$ *and* $\overline{f_\infty} = \infty$ *then there exists* $\lambda_0 > 0$ *such that problem (4.22) has two positive solutions if* $0 < \lambda < \lambda_0$.
5. *If* $\overline{f_0} = 0$ *and* $\overline{f_\infty} = \infty$ *then problem (4.22) has a positive solution for every* $\lambda > 0$.
6. *If* $\overline{f_0} = 0$ *and* $\overline{f_\infty} = 0$ *then there exists* $\lambda_0 > 0$ *such that problem (4.22) has two positive solutions if* $\lambda > \lambda_0$.
7. *If* $\overline{f_0} > 0$ *and* $\overline{f_\infty} > 0$ *then there exists* $\lambda_0 > 0$ *such that problem (4.22) has no positive solutions if* $\lambda > \lambda_0$.

Proof. Part 1. Fix $R > \gamma^* = 0$ and take $\lambda_0 = \lambda_0(R) > 0$ given by Lemma 49. In consequence, for all $0 < \lambda \leq \lambda_0(R)$ we have

$$\|\mathcal{T}_\lambda u\|_\infty \leq \|u\|_\infty \quad \text{for } u \in K \text{ with } \|u\|_\infty = R.$$

Now, let $0 < \lambda < \lambda_0$ be fixed, and choose $0 < r < \min\{R, r_0(\lambda)\}$, where $r_0(\lambda)$ is given by Lemma 52 when $\overline{f_0} = \infty$. In case of $\overline{f_\infty} = \infty$, we get $r > \max\{R, R_0(\lambda)\} > R$, with $R_0(\lambda)$ given by Lemma 54. In both situations we arrive at

$$\|\mathcal{T}_\lambda u\|_\infty \geq \|u\|_\infty \quad \text{for } u \in K \text{ with } \|u\|_\infty = r.$$

Thus Theorem 42 implies the existence of a positive solution for problem (4.22).

Part 2. Fix $r > 0$ and take $\lambda_0 = \lambda_0(r) > 0$ given by Lemma 50. Now, for each $\lambda > \lambda_0$ take $R > \max\{r, R_0(\lambda)\}$, with $R_0(\lambda)$ given by Lemma 53 and apply Theorem 42.

Part 3. For each $\lambda > 0$ take $r_0(\lambda) < R_0(\lambda)$ given by Lemmas 52 and 53, respectively, and apply Theorem 42.

Part 4. Fix $R_2 > R_1 > \gamma^* = 0$ and take $\lambda_0 = \min\{\lambda_0(R_1), \lambda_0(R_2)\}$ given by Lemma 49. Now for each $0 < \lambda < \lambda_0$ take $r < \min\{R_1, r_0(\lambda)\}$ given by Lemma 52 and $R > \max\{R_2, R_0(\lambda)\}$ given by Lemma 54. Then Theorem 42 implies the existence of two positive solutions u_1 and u_2 for problem (4.22) such that

$$r \leq \|u_1\| \leq R_1 < R_2 \leq \|u_2\| \leq R.$$

Part 5. Use Lemmas 51 and 54, and Theorem 42.

Part 6. Use Lemmas 50, 51, and 53 and Theorem 42 twice.

Part 7. The proof follows the same steps as Part 4 in Theorem 63.   □

Remark 44. Theorem 64 complements [41, Theorem 2.1], since it provides similar results for the problem $u''(t) + k^2 u(t) = \lambda g(t) f(u(t))$, with $0 < k < \pi/T$.

Example 14. Consider as a model the problem

$$u''(t) + e(1 + b \cos(t)) u(t) = \lambda \left(\frac{1}{u^\alpha(t)} + \mu u^\beta(t) \right),$$
$$u(0) = u(2\pi), u'(0) = u'(2\pi), \qquad\qquad (4.25)$$

where $e > 0$, $b \in \mathbb{R}$ and α, β, $\mu \geq 0$.

When $b = 1$ and $\mu = 0$ Eq. (4.25) is the Brillouin-beam focusing equation which has been widely studied in the literature (see [3,94,99] and references therein).

In Example 5 the linear part of the equation has been treated. There, the sufficient conditions (3.18) and (3.19) are obtained to ensure the positiveness of the related Green's function. As a consequence of those results and the previous one, we arrive at the following corollary.

Corollary 46. *Assuming condition (3.19), the following assertions hold*
(i) *If $0 \leq \beta < 1$ then problem (4.25) has a positive solution for every $\lambda > 0$.*
(ii) *If $\beta = 1$ and $\mu > 0$ then there exists $\lambda_0 > 0$ such that problem (4.25) has a positive solution for every $0 < \lambda < \lambda_0$ and there exists $\lambda_1 > 0$ such that the problem has no positive solution for $\lambda > \lambda_1$.*
(iii) *If $\beta > 1$ and $\mu > 0$ then there exists $\lambda_0 > 0$ such that problem (4.25) has two positive solutions for every $0 < \lambda < \lambda_0$.*

Proof. Condition (3.19) implies that condition (H1) is satisfied. Now, to prove (i), (ii) or (iii) it is enough to apply Theorem 64 Part 3, Part 1, and Part 7 or Part 4, respectively.   □

4.2.7.3 Non-negative Green's Function

In this section instead of conditions (H1) and (H3) we assume

$\widetilde{(H1)}$ Problem (4.2) is nonresonant, the corresponding Green's function

$$G_P[a] \text{ is non-negative on } I \times I \text{ and } \beta = \min_{t \in I} \int_0^T G_P[a](t, s)\, ds > 0,$$

$\widetilde{(H3)}$ $f : [0, \infty) \to [0, \infty)$ is continuous and $f(u) > 0$ for all $u > 0$.

Since

$$u(t) = \int_0^T G_P[a](t, s)\, ds$$

is the unique solution of the problem

$$u''(t) + a(t)\, u(t) = 1, \quad \text{a.e. } t \in I, \quad u(0) = u(T), \ u'(0) = u'(T),$$

assumption $\widetilde{(H1)}$ asks for this solution to be strictly positive on I. As a consequence, this condition is never fulfilled for Dirichlet or mixed boundary conditions.

On the other hand, assumption $\widetilde{(H3)}$ allows us to consider only regular problems. At the end of this subsection, we will turn out to singular problems by means of a truncation technique.

For a constant potential $a(t) \equiv k^2$, condition $\widetilde{(H1)}$ is equivalent to $0 < k \le \frac{\pi}{T}$. For a nonconstant potential a, Lemma 29 provides a sufficient condition that ensures that $\widetilde{(H1)}$ is fulfilled. Moreover, Lemma 24 ensures that condition $\widetilde{(H1)}$ is equivalent to

$$\lambda_0^P[a] < 0 \le \lambda_0^A[a].$$

On the other hand, under condition $\widetilde{(H1)}$ it is allowed that

$$m = \min_{t,s \in I} G_P[a](t, s) = 0,$$

so $\sigma = \dfrac{m}{M}$ can be equal to 0 and thus the arguments used in the previous section do not work. So, by assuming that $\gamma_* \ge 0$, let us define

$$\widetilde{K} := \left\{ u \in C(I), \ u \ge 0 \text{ on } I : \int_0^T u(s)\, ds \ge \tilde{\sigma}\, \|u\|_\infty \right\},$$

where $\tilde{\sigma} = \min\left\{ \dfrac{\beta}{TM}, \dfrac{\int_0^T \gamma(s)\, ds}{T \|\gamma\|} \right\}$ if $\|\gamma\| > 0$ or $\tilde{\sigma} = \dfrac{\beta}{TM}$ if $\|\gamma\| = 0$. As far as we know the cone $\widetilde{K}$ was introduced in [42].

Clearly $0 < \tilde{\sigma} \le 1$ and for $0 < r < R$ we define

$$\tilde{K}_{r,R} := \{u \in \tilde{K} : r \le \|u\|_\infty \le R\}.$$

Next, we prove the following result similar to Theorem 42.

Theorem 65. *Assume that (H0), $\widetilde{(H1)}$, (H2), $\widetilde{(H3)}$, and (H4) hold. Then, for each $\lambda > 0$ and $0 < r < R$ the operator $T_\lambda : \tilde{K}_{r,R} \to \tilde{K}$ given by (4.23) is well defined and completely continuous.*

Moreover, if either

(i) *$\|T_\lambda u\|_\infty \le \|u\|_\infty$ for any $u \in \tilde{K}$ with $\|u\|_\infty = r$ and $\|T_\lambda u\|_\infty \ge \|u\|_\infty$ for any $u \in \tilde{K}$ with $\|u\|_\infty = R$, or*

(ii) *$\|T_\lambda u\|_\infty \ge \|u\|_\infty$ for any $u \in \tilde{K}$ with $\|u\|_\infty = r$ and $\|T_\lambda u\|_\infty \le \|u\|_\infty$ for any $u \in \tilde{K}$ with $\|u\|_\infty = R$,*

then T_λ has a fixed point in $\tilde{K}_{r,R}$ which is a non-negative solution of problem (4.22).

Proof. Let $u \in C(I)$, $u \ge 0$ on I, and assuming $\|\gamma\| > 0$ (the case $\|\gamma\| = 0$ is analogous) we obtain

$$
\begin{aligned}
\int_0^T T_\lambda u(t)\, dt &= \lambda \int_0^T \int_0^T G_P[a](t,s)\, g(s) f(u(s))\, ds\, dt + \int_0^T \gamma(t)\, dt \\
&= \lambda \int_0^T g(s) f(u(s)) \left(\int_0^T G_P[a](t,s)\, dt \right) ds + \int_0^T \gamma(s)\, ds \\
&\ge \lambda \beta \int_0^T g(s) f(u(s))\, ds + \int_0^T \gamma(s)\, ds \\
&= \lambda \frac{\beta}{TM} TM \int_0^T g(s) f(u(s))\, ds + \frac{\int_0^T \gamma(s)\, ds}{T\|\gamma\|} T\|\gamma\| \\
&\ge \tilde{\sigma} \left(\lambda TM \int_0^T g(s) f(u(s))\, ds + T\|\gamma\| \right) \\
&\ge \tilde{\sigma} \|T_\lambda u\|_\infty.
\end{aligned}
$$

Thus, $T_\lambda(C(I)) \subset \tilde{K}$ and it is standard to show that T_λ is completely continuous. In consequence, from Krasnosel'skiĭ's fixed point Theorem (Theorem 42) it follows the existence of a fixed point of T_λ in $\tilde{K}_{r,R}$ which is, by definition of T_λ, a non-negative solution of problem (4.22). $\square$

Now we are going to give sufficient conditions to obtain $\|T_\lambda u\|_\infty \le \|u\|_\infty$ or $\|T_\lambda u\|_\infty \ge \|u\|_\infty$. The combination of the next lemmas with Theorem 65 will allow us to ensure the existence and multiplicity of solutions of problem (4.22).

Lemma 55. *Suppose that the conditions* (H0), $(\widetilde{H1})$, (H2), $(\widetilde{H3})$, *and* (H4) *are satisfied. Then, for each* $R > \gamma^*$ *there exists* $\lambda_0(R) > 0$ *such that for every* $0 < \lambda \le \lambda_0(R)$ *we have*

$$\|\mathcal{T}_\lambda u\|_\infty \le \|u\|_\infty \quad \text{for } u \in \widetilde{K} \text{ with } \|u\|_\infty = R.$$

Proof. Fix $R > \gamma^*$ and let $u \in \widetilde{K}$ with $\|u\|_\infty = R$. If

$$0 < \lambda \le \lambda_0(R) := \frac{R - \gamma^*}{M \max_{u \in [0,R]} f(u) \int_0^T g(s)\, ds}$$

then

$$\begin{aligned} \mathcal{T}_\lambda u(t) &= \lambda \int_0^T G_P[a](t,s)\, g(s)\, f(u(s))\, ds + \gamma(t) \\ &\le \lambda M \max_{u \in [0,R]} f(u) \int_0^T g(s)\, ds + \gamma^* \le R = \|u\|_\infty, \end{aligned}$$

and thus $\|\mathcal{T}_\lambda u\|_\infty \le \|u\|_\infty$. $\qquad\square$

Lemma 56. *Assume that* $(\widetilde{H1})$, (H2), $(\widetilde{H3})$, (H4), *and* $\int_0^T \gamma(s)\, ds > 0$ *are satisfied. Then there exists* $r_0 > 0$ *such that for each* $0 < r < r_0$ *we have*

$$\|\mathcal{T}_\lambda u\|_\infty \ge \|u\|_\infty \quad \text{for } u \in \widetilde{K} \text{ with } \|u\|_\infty = r.$$

Proof. Fix $0 < r < r_0 := \dfrac{1}{T} \int_0^T \gamma(s)\, ds$ and let $u \in \widetilde{K}$ with $\|u\|_\infty = r$. Then

$$\begin{aligned} \|\mathcal{T}_\lambda u\|_\infty &\ge \frac{1}{T} \int_0^T \mathcal{T}_\lambda u(t)\, dt \\ &= \frac{\lambda}{T} \int_0^T \int_0^T G_P[a](t,s)\, g(s)\, f(u(s))\, ds\, dt + \frac{1}{T} \int_0^T \gamma(t)\, dt \\ &\ge r_0 > r = \|u\|_\infty, \end{aligned}$$

and thus $\|\mathcal{T}_\lambda u\|_\infty \ge \|u\|_\infty$. $\qquad\square$

Lemma 57. *Let* (H0), $(\widetilde{H1})$, (H2), $(\widetilde{H3})$, *and* (H4) *be fulfilled. Then, if* $\overline{f_\infty} = 0$ *there exists* $R_0(\lambda) > 0$ *such that for every* $R \ge R_0(\lambda)$ *we have*

$$\|\mathcal{T}_\lambda u\|_\infty \le \|u\|_\infty \quad \text{for } u \in \widetilde{K} \text{ with } \|u\|_\infty = R.$$

Proof. Define $\tilde{f}(u) = \max\limits_{0 \leq z \leq u} f(z)$. Clearly $\tilde{f}$ is a monotone non-decreasing function on $[0, \infty)$ and moreover, since $\overline{f_\infty} = 0$ it is obvious that

$$\lim_{u \to \infty} \frac{\tilde{f}(u)}{u} = 0.$$

Therefore, we have that for $\varepsilon := \varepsilon(\lambda) = \frac{1}{2\lambda M \int_0^T g(s)\,ds}$ there exists $R_1(\lambda) > 0$ such that $\tilde{f}(u) \leq \varepsilon u$ for each $u \geq R_1(\lambda)$.

Define $R_0(\lambda) := \max\{R_1(\lambda), 2\gamma^*\}$, fix $R \geq R_0(\lambda)$ and let $u \in \tilde{K}$ with $\|u\|_\infty = R$. Then

$$\begin{aligned}
\mathcal{T}_\lambda u(t) &= \lambda \int_0^T G_P[a](t,s)\,g(s)f(u(s))\,ds + \gamma(t) \\
&\leq \lambda \int_0^T G_P[a](t,s)\,g(s)\tilde{f}(\|u\|_\infty)\,ds + \gamma(t) \\
&\leq \lambda M \varepsilon \|u\|_\infty \int_0^T g(s)\,ds + \gamma^* \\
&= \frac{R}{2} + \gamma^* \leq \frac{R}{2} + \frac{R}{2} = R = \|u\|_\infty,
\end{aligned}$$

and thus $\|\mathcal{T}_\lambda u\|_\infty \leq \|u\|_\infty$. $\qquad\square$

Theorem 66. *Assume* (H0), $(\widetilde{H1})$, (H2), $(\widetilde{H3})$, *and* (H4). *The following results hold:*
1. *If $\int_0^T \gamma(s)\,ds > 0$ then there exists $\lambda_0 > 0$ such that problem (4.22) has a non-negative solution if $0 < \lambda < \lambda_0$.*
2. *If $\int_0^T \gamma(s)\,ds > 0$ and $\overline{f_\infty} = 0$ then problem (4.22) has a non-negative solution for every $\lambda > 0$.*

Proof. The first assertion is a direct consequence of Lemmas 55 and 56. The second part follows from Lemmas 56 and 57. $\qquad\square$

Now we will impose a stronger condition on function g by assuming that it is strictly positive on the whole interval, that is,
$(\widetilde{H2})$ $g \in L^1(I)$, $g(t) \geq g_* > 0$ a.e. $t \in I$.

Lemma 58. *Assume that conditions* (H0), $(\widetilde{H1})$, $(\widetilde{H2})$, $(\widetilde{H3})$, *and* (H4) *are satisfied. Then, if $\overline{f_0} = \infty$ there exists $r_0(\lambda) > 0$ such that for every $0 < r \leq r_0(\lambda)$ we have*

$$\|\mathcal{T}_\lambda u\|_\infty \geq \|u\|_\infty \quad \text{for } u \in \tilde{K} \text{ with } \|u\|_\infty = r.$$

Proof. Since $\overline{f_0} = \infty$ for $L = L(\lambda) = \frac{T}{\lambda \beta \tilde{\sigma} g_*}$ there exists $r_0(\lambda) > 0$ such that $f(u) \geq Lu$ for each $0 \leq u \leq r_0(\lambda)$.

Fix $0 < r \leq r_0(\lambda)$ and let $u \in \tilde{K}$ with $\|u\|_\infty = r$. Then

$$
\begin{aligned}
\|\mathcal{T}_\lambda u\|_\infty &\geq \frac{1}{T} \int_0^T \mathcal{T}_\lambda u(t)\, dt \\
&= \frac{\lambda}{T} \int_0^T \int_0^T G_P[a](t,s)\, g(s) f(u(s))\, ds\, dt + \frac{1}{T} \int_0^T \gamma(t)\, dt \\
&\geq \frac{\lambda}{T} \int_0^T \int_0^T G_P[a](t,s)\, g(s) f(u(s))\, dt\, ds \geq \frac{\lambda}{T} \beta g_* L \int_0^T u(s)\, ds \\
&\geq \frac{\lambda}{T} \beta g_* L \tilde{\sigma} \|u\|_\infty = \|u\|_\infty,
\end{aligned}
$$

and thus $\|\mathcal{T}_\lambda u\|_\infty \geq \|u\|_\infty$. $\square$

Now, we are in a position to present the main result of this subsection.

Theorem 67. *Suppose that conditions* (H0), $(\widetilde{H1})$, $(\widetilde{H2})$, $(\widetilde{H3})$, *and* (H4) *are fulfilled. The following assertions are satisfied:*

1. *If* $\overline{f_0} = \infty$ *then there exists* $\lambda_0 > 0$ *such that problem (4.22) has a non-negative solution if* $0 < \lambda < \lambda_0$.
2. *If* $\overline{f_0} = \infty$ *and* $\overline{f_\infty} = 0$ *then problem (4.22) has a non-negative solution for every* $\lambda > 0$.
3. *If* $\underline{f_0} > 0$ *and* $\underline{f_\infty} > 0$ *then there exists* $\lambda_0 > 0$ *such that problem (4.22) has no non-negative solutions if* $\lambda > \lambda_0$.

Proof. The first assertion is a direct consequence of Lemmas 55 and 58. The second part follows from Lemmas 57 and 58 .

To prove Part 3, by using that $\underline{f_0} > 0$ and $\underline{f_\infty} > 0$ we know that there exists $L > 0$ such that $f(u) \geq Lu$ for all $u \geq 0$.

By defining

$$
\lambda_0 := \frac{T}{\beta \tilde{\sigma} L g_*},
$$

we have that if there is any $\lambda > \lambda_0$ for which there exists a non-negative solution u of problem (4.22), then $u = \mathcal{T}_\lambda u \in K$. So, we arrive at the following contradiction

$$
\begin{aligned}
\|\mathcal{T}_\lambda u\|_\infty &\geq \frac{1}{T} \int_0^T \mathcal{T}_\lambda u(t)\, dt \\
&= \frac{\lambda}{T} \int_0^T \int_0^T G_P[a](t,s)\, g(s) f(u(s))\, ds\, dt + \frac{1}{T} \int_0^T \gamma(t)\, dt
\end{aligned}
$$

$$\geq \frac{\lambda}{T}\int_0^T\int_0^T G_P[a](t,s)\,g(s)f(u(s))\,dt\,ds \geq \frac{\lambda}{T}\,\beta\,g_*\,L\int_0^T u(s)\,ds$$

$$\geq \frac{\lambda}{T}\,\beta\,g_*\,L\widetilde{\sigma}\,\|u\|_\infty > \|u\|_\infty. \quad \square$$

Applications to Singular Equations

Despite in the previous results we deal with regular functions, it is possible to apply some of them to the singular equation

$$u''(t) + a(t)\,u(t) = \lambda\,g(t)f(u(t)) + c(t), \quad \text{a.e. } t \in I,$$
$$u(0) = u(T), \quad u'(0) = u'(T), \tag{4.26}$$

by means of a truncation technique.

To this end, we will consider a function f that satisfies

(H5) $f : (0,\infty) \to (0,\infty)$ is a continuous function such that $\overline{f_\infty} = 0$.

Theorem 68. *Assume $\gamma_* > 0$ and conditions $(\widetilde{H1})$, (H2), (H4), and (H5). Then problem (4.26) has a positive solution for every $\lambda > 0$.*

Proof. Let $r = \gamma_* > 0$ and define the function

$$f_r(u) = \begin{cases} f(r), & \text{if } 0 \leq u < r, \\ f(u), & \text{if } u \geq r. \end{cases}$$

From (H5) it follows that f_r satisfies condition $(\widetilde{H3})$ and $\overline{f_{r\infty}} = 0$. Moreover $\gamma_* > 0$ implies that $\int_0^T \gamma(s)ds > 0$. As a consequence Theorem 66, Part 2, implies that the modified problem

$$u''(t) + a(t)\,u(t) = \lambda\,g(t)f_r(u(t)) + c(t), \quad \text{a.e. } t \in I,$$
$$u(0) = u(T), \quad u'(0) = u'(T),$$

has a non–negative solution u_r for all $\lambda > 0$. Such function is given by the expression

$$u_r(t) = \lambda\int_0^T G_P[a](t,s)\,g(s)f_r(u_r(s))\,ds + \gamma(t).$$

The non–negativeness of functions $G_P[a]$, g and f_r implies that the solution $u_r(t) \geq \gamma_* = r$ for all $t \in I$. Therefore u_r is a positive solution of problem (4.26). $\square$

Remark 45. Theorem 68 is an alternative result to those obtained in [27, 95] by means of Schauder's fixed point Theorem.

Example 15. Let us consider the repulsive singular differential equation

$$u''(t) + a(t)\, u(t) = \lambda\, g(t) \left(\sqrt{u(t)} - \log(u(t)) \right) + c(t). \tag{4.27}$$

Since $f(x) = \sqrt{x} - \log(x)$ satisfies $(H5)$, we can apply Theorem 68 to obtain the following result.

Corollary 47. *Assume $(\widetilde{H1})$, $(H2)$, and $(H4)$. If $\gamma_* > 0$ then Eq. (4.27) has a positive T-periodic solution for every $\lambda > 0$.*

4.3 LOWER AND UPPER SOLUTIONS METHOD

The method of lower and upper solutions is a classical tool in the theory of nonlinear boundary value problems. It allows us to ensure the existence of a solution of the considered problem under the assumption of existence of a pair of functions that satisfy some suitable inequalities. Such functions are known as lower and upper solutions. When they are well ordered, the found solution lies between both of them. So, in this particular case, we have information not only about the existence of solutions but also about the location of some of them. On the contrary, if the lower and upper solutions appear either in the reverse order or without any one, under some additional conditions on the nonlinear part of the equation, it is possible to ensure the existence of solution but, in this case, it can only be given some information about its location at some subinterval of definition. Unfortunately, there is not a direct way of constructing the pair of functions we are looking for. It is because of this that along the literature there have been developed different generalizations on the definitions of lower and upper solutions, that allow drop some regularity conditions and, as a consequence, enlarge the set of possible situations where it is possible to apply this method.

In this section we will concentrate on the deep dependence of the validity of this method and the constant sign of the related Green's function. So, we will impose to the lower and upper solutions the same regularity as the solutions we are looking for. We will indicate some references where the reader can look for additional information about this theory in a wider framework.

We start by making a small survey of this theory which is compiled in [7]. The reader can also consult the monographs [8,34].

The first steps in the theory of lower and upper solutions were given by E. Picard in 1890 [87] for partial differential equations and, three years after, in [86] for ordinary differential equations. In both cases the existence of a solution is guaranteed from a monotone iterative technique. Existence of solutions for Cauchy equations have been proved by O. Perron in 1915 [83]. In 1926 Müller extended Perron's results to initial value systems in [77].

G. Scorza Dragoni [92] introduced in 1931 the notion of lower and upper solutions for ordinary differential equations with Dirichlet boundary value conditions. In particular the author considered the second order boundary value problem (on a general interval $[a, b]$)

$$u''(t) = f(t, u(t), u'(t)), \ t \in I, \quad u(0) = A, \ u(T) = B, \tag{4.28}$$

for $f : I \times \mathbb{R}^2 \to \mathbb{R}$ a continuous function and $A, B \in \mathbb{R}$.

A lower solution is a function $\alpha \in C^2(I)$ that satisfies the inequality

$$\alpha''(t) \geq f(t, \alpha(t), \alpha'(t)), \quad t \in I, \tag{4.29}$$

together with

$$\alpha(0) \leq A, \ \alpha(T) \leq B.$$

In the same way, an upper solution is a function $\beta \in C^2(I)$ that satisfies the reversed inequalities

$$\beta''(t) \leq f(t, \beta(t), \beta'(t)), \quad t \in I, \tag{4.30}$$

and

$$\beta(0) \geq A, \ \beta(T) \geq B.$$

When $\alpha \leq \beta$ on I, the existence of a solution of the considered problem lying between α and β was proved. So the problem of finding a solution of the considered problem is replaced by that of finding two well ordered functions that satisfy some suitable inequalities.

Following these pioneering results, there have been a large number of works in which the method has been developed for different kinds of boundary value problems, thus first, second, and higher order ordinary differential equations with different types of boundary conditions such as, among others, periodic, mixed, Dirichlet or Neumann, have been considered. Also partial differential equations, of first and second order, have been treated in the literature.

In these situations, we have that for the Neumann problem

$$u''(t) = f(t, u(t), u'(t)), \ t \in I, \quad u'(0) = A, \ u'(T) = B, \tag{4.31}$$

a lower solution $\alpha \in C^2(I)$ is a function that satisfies (4.29) coupled with the inequalities

$$\alpha'(0) \geq A, \ \alpha'(T) \leq B.$$

$\beta \in C^2(I)$ is an upper solution of the Neumann problem if it satisfies (4.30) and

$$\beta'(0) \leq A, \ \beta'(T) \geq B.$$

Analogously, for the periodic problem

$$u''(t) = f(t, u(t), u'(t)), \ t \in I, \quad u(0) = u(T), \ u'(0) = u'(T), \tag{4.32}$$

a lower solution α and an upper solution β are C^2-functions that satisfy (4.29) and (4.30), respectively, together with the inequalities

$$\alpha(0) = \alpha(T), \ \alpha'(0) \geq \alpha'(T)$$

and

$$\beta(0) = \beta(T), \ \beta'(0) \leq \beta'(T).$$

Note that the linear part of these two problems is resonant, so they are not covered by the results concerning Eq. (4.1).

In the well-known monograph of S.R. Bernfeld and V. Lakshmikantham [4] the classical theory of lower and upper solutions is developed. We refer the reader to the classical works of J. Mawhin [71–74] and the surveys in this field of A. Cabada [7], and C. De Coster and P. Habets [31,34] and their monograph [33], in which one can find historical and bibliographical references together with recent results and open problems.

It is important to point out that, to derive the existence of a solution, a growth condition on the nonlinear part of the equation with respect to the dependence on the first derivative is imposed. The most usual condition is the so-called Nagumo's condition that was introduced by this author in 1937 [78]. This condition imposes, roughly speaking, a quadratic growth in the dependence of the derivative. The most common form of presenting it is the following one.

Definition 19. We say that $f : I \times \mathbb{R}^2 \longrightarrow \mathbb{R}$ satisfies the Nagumo's condition related to a pair of well ordered functions $\alpha \le \beta$ on I, if for all $(t, x, y) \in I \times [\alpha(t), \beta(t)] \times \mathbb{R}$ there is $h \in C(I)$ satisfying

$$|f(t, x, y)| \le h(|y|) \tag{4.33}$$

and

$$\int_v^\infty \frac{s\, ds}{h(s)} = \infty, \tag{4.34}$$

with

$$v = \frac{\max\{|\beta(T) - \alpha(0)|, |\beta(0) - \alpha(T)|\}}{T}. \tag{4.35}$$

The main importance of this condition is that it provides a priori bounds on the first derivative of all the possible solutions of the studied problem that lie between the lower and the upper solution. A careful proof of this property has been made in [4]. One can verify that in the proof the condition (4.34) can be replaced by the weaker one:
There exists a real constant $K > 0$ such that

$$\int_v^K \frac{s\, ds}{h(s)} > \max_{t \in I} \beta(t) - \min_{t \in I} \alpha(t).$$

The usual tool to derive an existence result consists in the construction of a modified problem that satisfies the two following properties:
1. The nonlinear part of the modified equation is bounded.
2. The nonlinear part of the modified equation coincides with the nonlinear part when the spatial variable is in $[\alpha, \beta]$.

4.3.1 Well Ordered Lower and Upper Solutions

To be concise, we will develop this method for the periodic problem (4.32). So, we will follow the arguments done by W. Gao and J. Wang in [40], which improve the previous ones given by M.X. Wang, A. Cabada, and J.J. Nieto in [97]. As we can see, the arguments hold, with simple technical adaptations, to other linear boundary conditions, as the Neuman or the Dirichlet ones.

As in the previous section, we will work with f a Carathéodory function and we will look for solutions on $W^{2,1}(I)$. As a consequence Eq. (4.32) is fulfilled a.e. $t \in I$.

This regularity makes necessary to modify the definition of a lower and an upper solution for this new situation. So, we have the following definition.

Definition 20. We say that $\alpha \in W^{2,1}(I)$ is a lower solution of problem (4.32), if it satisfies the following inequalities:

$$\alpha''(t) \geq f(t, \alpha(t), \alpha'(t)), \quad \text{a.e.} \quad t \in I, \quad \alpha(0) = \alpha(T), \; \alpha'(0) \geq \alpha'(T).$$

$$(4.36)$$

Analogously, $\beta \in W^{2,1}(I)$ is an upper solution of problem (4.32) if the reversed inequalities are fulfilled.

Next we define the Nagumo's condition we are going to use, which is used for ϕ-Laplacian equations and nonlinear boundary value conditions in [22]. Note that the condition does not depend on the boundary data of the problem. So, it can be used for any other, linear or nonlinear, boundary condition.

Definition 21. We say that $f : I \times \mathbb{R}^2 \to \mathbb{R}$, a Carathéodory function, satisfies a Nagumo's condition relative to the pair α and β, with α, $\beta \in C(I)$, $\alpha \leq \beta$ in I, if there exist functions $k \in L^p(I)$, $1 \leq p \leq \infty$, and $\theta : [0, \infty) \longrightarrow (0, \infty)$ continuous, such that

$$|f(t, u, v)| \leq k(t)\theta(|v|) \text{ a.e. } (t, u, v) \in \Omega,$$

where $\Omega = \{(t, u, v) \in I \times \mathbb{R}^2 : \alpha(t) \leq u \leq \beta(t)\}$, and also that

$$\min\left\{\int_v^\infty \frac{|u|^{\frac{p-1}{p}}}{\theta(|u|)} du, \int_{-\infty}^{-v} \frac{|u|^{\frac{p-1}{p}}}{\theta(|u|)} du\right\} > \mu^{\frac{p-1}{p}} \|k\|_p,$$

with v given in Definition 19,

$$\mu = \max_{t \in I} \beta(t) - \min_{t \in I} \alpha(t),$$

and

$$\|k\|_p = \begin{cases} \sup_{t \in I} |k(t)| & \text{if} \quad p = \infty, \\ \left(\int_0^T |k(t)|^p dt\right)^{\frac{1}{p}} & \text{if} \quad 1 \leq p < \infty, \end{cases}$$

where $(p-1)/p \equiv 1$ for $p = \infty$.

Remark 46. It is clear that in Definitions 19 and 21 we have that

$$v = \frac{\beta(0) - \alpha(0)}{T}.$$

It has been used the more general definition (4.35) because it covers any kind of boundary conditions and not only the periodic ones.

The main result of this subsection is the following existence and location theorem for the periodic problem (4.32).

Theorem 69. *Let α and β be a lower and an upper solution respectively for the periodic boundary value problem (4.32) such that $\alpha \le \beta$ in I. Assume that f is a Carathéodory function that satisfies the Nagumo's condition relative to α and β introduced in Definition 21.*

Then problem (4.32) has, at least, a solution $u \in [\alpha, \beta]$. Moreover it satisfies that $M_- < u'(t) < M_+$ for all $t \in I$, where M_+ and M_- are two real constants that only depend on $\alpha, \beta, \theta,$ and k.

In order to prove the previous existence theorem, we need to introduce some definitions and prove some preliminary lemmas.

First, we define

$$p(t, x) = \max\{\alpha(t), \min\{x, \beta(t)\}\} \quad \text{for all } t \in I \text{ and } x \in \mathbb{R}. \tag{4.37}$$

Obviously, due to the continuity of α and β, we have that for any continuous function u, the composed function $p(\cdot, u(\cdot))$ is continuous too. Of course, at the points where they attain the same value, but with a different slope, the new function is not C^1. However, it satisfies the following regularity property.

Lemma 59. *[97, Lemma 2] For each $u \in C^1(I)$ the following properties hold:*

(i) $\dfrac{d}{dt}p(t, u(t))$ *exists a.e. $t \in I$.*

(ii) *If $u, u_m \in C^1(I)$ and u_m converges in $C^1(I)$ to u, then*

$$\lim_{m \to \infty} \left\{ \frac{d}{dt} p(t, u_m(t)) \right\} = \frac{d}{dt} p(t, u(t)) \quad \text{a.e. } t \in I.$$

Proof. First, note that if $u \in C^1(I)$ then $u^+ = \max\{u, 0\}$ and $u^- = \max\{-u, 0\}$ are absolutely continuous functions on I. As a consequence, both $\dfrac{d}{dt}u_m^+(t)$ and $\dfrac{d}{dt}u^+(t)$ exist a.e. $t \in I$.

We rewrite

$$p(t, u(t)) = [u(t) - \alpha(t)]^- - [u(t) - \beta(t)]^+ + u(t), \quad t \in I, \qquad (4.38)$$

which says us that $p(\cdot, u(\cdot)) \in AC(I)$ and assertion (i) is fulfilled.

Since $u, \alpha, \beta \in C^1(I)$, from (4.38), it is enough to prove that if $u, u_m \in C^1(I)$ and u_m converges in $C^1(I)$ to u, then

$$\lim_{m \to \infty} \left\{ \frac{d}{dt} u_m^{\pm}(t) \right\} = \frac{d}{dt} u^{\pm}(t) \quad \text{a.e. } t \in I.$$

Choose $t_0 \in I$ such that $\frac{d}{dt} u_m^+(t_0)$ and $\frac{d}{dt} u^+(t_0)$ exist for all $m \in \mathbb{N}$.

If $u(t_0) > 0$, then $u(t_0) = u^+(t_0) > 0$. Therefore $\frac{d}{dt} u^+(t_0) = \frac{d}{dt} u(t_0)$ and there exists $m_0 \in \mathbb{N}$ such that $u_m(t_0) = u_m^+(t_0) > 0$ for all $m \geq m_0$. Thus

$$\lim_{m \to \infty} \left\{ \frac{d}{dt} u_m^+(t_0) \right\} = \lim_{m \to \infty} \left\{ \frac{d}{dt} u_m(t_0) \right\} = \frac{d}{dt} u(t_0).$$

When $u(t_0) < 0$ there exists $m_1 \in \mathbb{N}$ such that $u_m(t_0) < 0$ for all $m \geq m_1$. Therefore $u_m(t) < 0$ on $(t_0 - \delta_m, t_0 + \delta_m)$ for some $\delta_m > 0$ and then $u_m^+(t) = 0$ on $(t_0 - \delta_m, t_0 + \delta_m)$.

Hence $\frac{d}{dt} u_m^+(t_0) = \frac{d}{dt} u^+(t_0) = 0$ and then

$$\lim_{m \to \infty} \left\{ \frac{d}{dt} u_m^+(t_0) \right\} = \lim_{m \to \infty} \left\{ \frac{d}{dt} u^+(t_0) \right\} = 0.$$

If $u(t_0) = 0$, then $u^+(t_0) = 0$. Since $\frac{d}{dt} u^+(t_0)$ exists, we have that $\frac{d}{dt} u^+(t_0) = 0$. Now, due to the C^1 regularity of function u, we deduce that $\frac{d}{dt} u(t_0) = 0$.

As a consequence, since $\frac{d}{dt} u_m^+(t_0)$ exists, we find that

$$\frac{d}{dt} u_m^+(t_0) = \begin{cases} u_m'(t_0) & \text{if } u_m(t_0) > 0, \\ 0 & \text{if } u_m(t_0) \leq 0. \end{cases}$$

Therefore, due to the convergence in $C^1(I)$, we have

$$0 \leq \lim_{m \to \infty} \left\{ \left| \frac{d}{dt} u_m^+(t_0) \right| \right\} \leq \lim_{m \to \infty} \left\{ \left| \frac{d}{dt} u_m(t_0) \right| \right\} = 0 = \frac{d}{dt} u^+(t_0).$$

The same arguments hold for u^- and the proof is completed. $\square$

Construction of the modified problem

In order to construct a truncated problem related to (4.32) that satisfies the two suitable mentioned properties of boundedness and coincidence in $[\alpha, \beta]$, we use the fact that, by Definition 21, we can find two real numbers, $M_- < 0 < M_+$, such that $M_- < -\nu \leq \nu < M_+$, $M_- < \alpha'(t)$, $\beta'(t) < M_+$ for all $t \in I$, and

$$\min \left\{ \int_\nu^{M_+} \frac{|u|^{\frac{p-1}{p}}}{\theta(|u|)} du, \int_{M_-}^{-\nu} \frac{|u|^{\frac{p-1}{p}}}{\theta(|u|)} du \right\} > \mu^{\frac{p-1}{p}} \|k\|_p.$$

We note that M_+ and M_- only depend on α, β, θ, and k. So, for all $y \in \mathbb{R}$, we can define the function

$$q(y) = \max \{M_-, \min \{y, M_+\}\}, \tag{4.39}$$

and consider the following modified problem

$$u''(t) = f\left(t, p(t, u(t)), q\left(\frac{d}{dt} p(t, u(t))\right)\right) \quad \text{a.e. } t \in I, \tag{4.40}$$

$$u(0) = p(0, u(0) + u'(0) - u'(T)), \tag{4.41}$$

$$u(T) = p(0, u(0)). \tag{4.42}$$

We point out that, as it has been noted in Lemma 59, u^+ and u^- are absolutely continuous functions on I for any $u \in C^1(I)$. As a consequence, from (4.38) we deduce that $p(\cdot, u(\cdot))$ is an absolutely continuous function on I too. So, from (4.39), we know that $q\left(\frac{d}{dt} p(\cdot, u(\cdot))\right)$ belongs to $L^\infty(I)$. As a conclusion, we have that, provided that f is a Carathéodory function, the right hand side of the previous equation is in $L^1(I)$ for all $u \in C^1(I)$.

Now, denoting as

$$[\alpha, \beta] = \{u \in C^1(I); \ \alpha(t) \leq u(t) \leq \beta(t), \text{ for all } t \in I\},$$

and, in order to ensure the existence of solution of the periodic problem (4.32), we prove the following lemmas for the truncated problem (4.40)–(4.42).

Lemma 60. *If u is a solution of (4.40)–(4.42) then $u \in [\alpha, \beta]$ and $u(0) = u(T)$.*

Proof. We shall only see that $\alpha(t) \leq u(t)$ for every $t \in I$. An analogous reasoning shows that $u(t) \leq \beta(t)$ for all $t \in I$.

By definition of p, using (4.41), we have that $\alpha(0) \leq u(0) \leq \beta(0)$.

Moreover, using (4.42), by the definition of lower and upper solutions, we deduce that

$$\alpha(T) = \alpha(0) \leq u(T) \leq \beta(0) = \beta(T).$$

Assume, on the contrary, that

$$\min_{t \in I} \{(u - \alpha)(t)\} < 0,$$

then, since $u - \alpha \in C^1(I)$, we have that there exist $0 \leq t_1 < t_2 \leq T$ such that $u < \alpha$ in (t_1, t_2) and $(u - \alpha)(t_1) = (u - \alpha)(t_2) = 0$.

Thus, from the definition of α, p and q, we have that there is $h_\alpha \in L^1(I)$, $h_\alpha \geq 0$ a.e. on I, such that

$$(u - \alpha)''(t) = f(t, \alpha(t), \alpha'(t)) - f(t, \alpha(t), \alpha'(t)) - h_\alpha(t), \qquad \text{a.e. } t \in (t_1, t_2).$$

In consequence, we have that $u - \alpha$ solves the following linear Dirichlet problem on the interval (t_1, t_2):

$$-w''(t) = h_\alpha(t), \quad \text{a.e. } t \in (t_1, t_2), \quad w(t_1) = w(t_2) = 0. \tag{4.43}$$

But, as we have seen in Chapter 3, $\lambda = 0$ is not an eigenvalue of previous operator at any real interval. As a consequence, problem (4.43) has a unique solution given by

$$w(t) = -\int_{t_1}^{t_2} G_D[0](t, s) \, h_\alpha(s) \, ds, \quad t \in [t_1, t_2],$$

where the Green's function $G_D[0]$, see [8], is given by the expression

$$G_D[0](t, s) = -\frac{1}{t_2 - t_1} \begin{cases} (s - t_1)(t_2 - t), & t_1 \leq s \leq t \leq t_2, \\ (t_2 - s)(t - t_1), & t_2 < t \leq s \leq t_2. \end{cases}$$

It is immediate to verify that $-G_D[0](t, s) > 0$ for all $(t, s) \in (t_1, t_2) \times (t_1, t_2)$. So, from the non-negativeness of function h_α, we deduce that $w \geq 0$ on (t_1, t_2), that is, $u \geq \alpha$ on this interval and we arrive at a contradiction.

Now, (4.42) reads as follows:

$$u(T) = p(0, u(0)) = u(0). \qquad \qquad \square$$

Lemma 61. *If u is a solution of (4.40)–(4.42) then $M_- < u'(t) < M_+$ for all $t \in I$, where M_+ and M_- are the Nagumo constants that were chosen for the construction of (4.40)–(4.42), and only depend on α, β, θ, and k.*

Proof. Let $u \in C^1(I)$ be a solution of (4.40)–(4.42). As we know, by Lemma 60, $u \in [\alpha, \beta]$, and then

$$u''(t) = f(t, u(t), q(u'(t))) \text{ a.e. } t \in I, \quad u(0) = u(T).$$

Thus, by the mean value theorem, there exists $t_0 \in (0, T)$ such that $u'(t_0) = 0$, and then

$$M_- < -\nu \le \frac{\alpha(0) - \beta(0)}{T} \le u'(t_0) \le \frac{\beta(0) - \alpha(0)}{T} \le \nu < M_+.$$

Suppose that there exists a point in the interval I for which $u' > M_+$ or $u' < M_-$. By the continuity of u' we can choose $t_1 \in I$ satisfying one of the following conditions

(i) $u'(t_0) = 0$, $u'(t_1) = M_+$ and $0 \le u'(t) \le M_+$ for all $t \in (t_0, t_1)$,
(ii) $u'(t_1) = M_+$, $u'(t_0) = 0$ and $0 \le u'(t) \le M_+$ for all $t \in (t_1, t_0)$,
(iii) $u'(t_0) = 0$, $u'(t_1) = M_-$ and $M_- \le u'(t) \le 0$ for all $t \in (t_0, t_1)$,
(iv) $u'(t_1) = M_-$, $u'(t_0) = 0$ and $M_- \le u'(t) \le 0$ for all $t \in (t_1, t_0)$.

Assume that the situation (i) holds (the other cases admit analogous proofs).

Since $M_- \le 0 \le u'(t) \le M_+$ for all $t \in (t_0, t_1)$, we have

$$u''(t) = f(t, u(t), q(u'(t))) = f(t, u(t), u'(t)) \text{ a.e. } t \in (t_0, t_1),$$

so, by the Nagumo's condition,

$$|u''(t)| = |f(t, u(t), u'(t))| \le k(t)\theta(|u'(t)|) \text{ a.e. } t \in (t_0, t_1). \quad (4.44)$$

Thus, the fact that $0 \le \nu$ and the positiveness of function θ leads us to

$$\int_0^{M_+} \frac{s^{\frac{p-1}{p}}}{\theta(s)}\, ds \ge \int_\nu^{M_+} \frac{s^{\frac{p-1}{p}}}{\theta(s)}\, ds > \mu^{\frac{p-1}{p}}\|k\|_p. \quad (4.45)$$

Now, by means of the change of variable $s = u'(t) \ge 0$, we have (see [75]) that

$$\int_0^{M_+} \frac{s^{\frac{p-1}{p}}}{\theta(s)}\, ds = \int_{t_0}^{t_1} \frac{(u'(t))^{\frac{p-1}{p}} u''(t)}{\theta(u'(t))}\, dt \le \int_{t_0}^{t_1} \frac{|u'(t)|^{\frac{p-1}{p}} |u''(t)|}{\theta(u'(t))}\, dt.$$

Thus, using (4.44) and the fact that $u' \geq 0$ on (t_0, t_1), we have, by means of Hölder's inequality, that

$$\int_0^{M_+} \frac{s^{\frac{p-1}{p}}}{\theta(s)} \, ds \leq \int_{t_0}^{t_1} k(t)(u'(t))^{\frac{p-1}{p}} \, dt \leq \|k\|_p \, \mu^{\frac{p-1}{p}},$$

which is a contradiction with (4.45).   □

Lemma 62. *If u is a solution of (4.40)–(4.42) then $u'(0) = u'(T)$.*

Proof. If we show that $\alpha(0) \leq u(0) + u'(0) - u'(T) \leq \beta(0)$, then (4.41) implies that $u(0) = u(0) + u'(0) - u'(T)$, and the result holds.

Suppose, on the contrary, that $u(0) + u'(0) - u'(T) < \alpha(0)$. Using (4.41), we obtain

$$u(0) = p(0, u(0) + u'(0) - u'(T)) = \alpha(0)$$

and then, from (4.42) and the definition of α, we have

$$u(T) = \alpha(T).$$

Then, since $u - \alpha \in C^1(I)$, $(u - \alpha)(0) = (u - \alpha)(T) = 0$, and $u \geq \alpha$ in I (Lemma 60), we have that $(u - \alpha)'(0) \geq 0 \geq (u - \alpha)'(T)$.

Thus, using again the definition of α, we conclude that

$$u(0) + u'(0) - u'(T) \geq \alpha(0) + \alpha'(0) - \alpha'(T) \geq \alpha(0),$$

reaching a contradiction.

In a similar way, one can verify that $u(0) + u'(0) - u'(T) \leq \beta(0)$ and the result is concluded.   □

By Lemmas 60, 61, and 62, the proof of the existence result Theorem 69 will be over once we show that problem (4.40)–(4.42) admits a solution.

To this end, we construct an operator whose fixed points coincide with the solutions of such problem. Since the modified problem is a kind of a nonhomogeneous Dirichlet problem, it is immediate to verify that the solutions of the modified problem coincide with the fixed points of the operator

$$\mathcal{T} : C^1(I) \to C^1(I),$$

defined, for all $t \in I$, as

$$\mathcal{T}u(t) = \int_0^T G_D[0](t, s) f\left(s, p(s, u(s)), q\left(\frac{d}{ds} p(s, u(s))\right)\right) ds$$

$$+\frac{T-t}{T}p(0,u(0))+u'(0)-u'(T))+\frac{t}{T}p(0,u(0)).$$

Here, the Green's function $G_D[0]$ is the related one to the Dirichlet problem on the interval I:

$$u''(t)=0,\ t\in I,\quad u(0)=u(T)=0,$$

which is given by the following expression:

$$G_D[0](t,s)=-\frac{1}{T}\begin{cases} s(T-t), & 0\le s\le t\le T, \\ t(T-s), & 0<t<s\le T. \end{cases}\qquad (4.46)$$

Obviously, we have that

$$\begin{aligned}(T\,u)'(t) &= \frac{1}{T}\int_0^t sf\left(s,p(s,u(s)),q\left(\frac{d}{ds}p(s,u(s))\right)\right)ds \\ &+\frac{1}{T}\int_t^T (s-T)f\left(s,p(s,u(s)),q\left(\frac{d}{ds}p(s,u(s))\right)\right)ds \\ &-\frac{1}{T}p(0,u(0))+u'(0)-u'(T))+\frac{1}{T}p(0,u(0)).\end{aligned}$$

To deduce the solvability of the modified problem (4.40)–(4.42), we use the Schaefer's fixed point Theorem (Theorem 37). To this end, we need to ensure the regularity of the constructed operator T.

Lemma 63. *Operator T is completely continuous.*

Proof. Using the dominated Lebesgue convergence Theorem, from Lemma 59, we have that T is a continuous operator.

From the definition of functions p, q, it is obvious that there is $h\in L^1(I)$ such that

$$\left|f\left(t,p(t,u(t)),q\left(\frac{d}{dt}p(t,u(t))\right)\right)\right|\le h(t),\quad \text{for a.e. } t\in I \text{ and all } u\in C^1(I).$$

So, by the definition of $G_D[0]$, we conclude that there is $R>0$ such that

$$\max\{\|T\,u\|_\infty, \|(T\,u)'\|_\infty\}\le R,\quad \text{for all } u\in C^1(I).$$

In particular, we have that T maps bounded sets of $C^1(I)$ into bounded sets of $C^1(I)$.

Let's see that $\mathcal{T}(M)$ is, actually, a relatively compact set, whenever M is bounded. By Ascoli–Arzelà's Theorem, this property is equivalent to verify that the set $\mathcal{T}(M)$ is equicontinuous in $C^1(I)$.

Thus, for all $u \in C^1(I)$ we have that

$$|\mathcal{T}u(t_2) - \mathcal{T}u(t_1)| \leq \int_{t_1}^{t_2} |(\mathcal{T}u)'(s)|\, ds \leq |t_2 - t_1|\, R$$

and

$$|(\mathcal{T}u)'(t_2) - (\mathcal{T}u)'(t_1)| = \left| \int_{t_1}^{t_2} f\left(s, p(s, u(s)), q\left(\frac{d}{ds}p(s, u(s))\right)\right)\, ds \right|$$
$$\leq |t_2 - t_1|\, \|h\|_{L^1(I)}.$$

So, we conclude that $\mathcal{T}(M)$ is an equicontinuous set on $C^1(I)$.   □

As a consequence, if $u = \lambda\,\mathcal{T}u$ for some $\lambda \in [0, 1)$, then

$$\max\{\|u\|_\infty, \|u'\|_\infty\} \leq \lambda\, R < R.$$

Thus, Schaefer's fixed point Theorem (Theorem 37) holds and Theorem 69 is proved.

Remark 47. As it is pointed out in the proof of Theorem 69, the main used arguments do not depend on the boundary data of the considered problem. So, this result is valid for a wider set of boundary conditions, including nonlinear and functional ones (see [21,23] and references therein). In particular it is true for the Dirichlet conditions $u(0) = A$, $u(T) = B$ (in this case assuming $\alpha(0) \leq A \leq \beta(0)$ and $\alpha(T) \leq B \leq \beta(T)$) and the Neumann ones $u'(0) = A$, $u'(T) = B$ (imposing $\alpha'(0) \geq A \geq \beta'(0)$ and $\alpha'(T) \leq B \leq \beta'(T)$).

We note that (see also [8,34]) contrary to the truncated problem (4.40)–(4.42), not all the solutions of problem (4.32) must be in $[\alpha, \beta]$. To see this, it is enough to consider the problem

$$u''(t) = u(t)\,(u(t) - 1),\ t \in I, \quad u(0) = u(T),\ u'(0) = u'(T),$$

for which $\alpha \equiv 1/2$ and $\beta \equiv 3/2$ are a pair of well ordered lower and upper solutions.

It is obvious that $u \equiv 1$ is a solution of this problem and it is on $[\alpha, \beta]$. However, $v \equiv 0$ is another solution that, obviously, does not belong to $[\alpha, \beta]$.

It is important to point out, see [7,8,34] and references therein, that in 1954, M. Nagumo constructed an example, [79], in which the existence of well-ordered lower and upper solutions, without assuming the Nagumo's condition, is not sufficient to ensure the existence of solutions of a Dirichlet problem, i.e., in general this growth condition cannot be removed for the Dirichlet case. An analogous result concerning the optimality of the Nagumo's condition for periodic and Sturm-Liouville conditions was showed in 2003 by P. Habets and R.L. Pouso in [45].

In 1967 I.T. Kiguradze [55] proved that it is enough to consider a one sided Nagumo condition (by eliminating the absolute value in (4.33)) to deduce existence results for Dirichlet problems. Similar results were given in 1968 by K.W. Schrader [90].

It is important to note that there are many papers that have tried to get existence results under weaker assumptions on the definition of lower and upper solutions. In particular G. Scorza Dragoni proved in 1938 [93], an existence result for the Dirichlet problem by assuming the existence of two C^1 functions $\alpha \leq \beta$ such that

$$\alpha'(t) - \int^t f(s, \alpha(s), \alpha'(s))\, ds$$

and

$$-\beta'(t) + \int^t f(s, \beta(s), \beta'(s))\, ds$$

are monotone non-decreasing in $t \in I$.

I.T. Kiguradze used in [56] regular lower and upper solutions and explained that it is possible to get the same results for lower and upper solutions whose first derivatives are not absolutely continuous functions. V.D. Ponomarev considered in [88] two continuous functions $\alpha, \beta : I \to \mathbb{R}$ with right Dini derivatives $D_r\alpha$, absolute semi-continuous from below in I, and $D_r\beta$ absolute semi-continuous from above in I, that satisfy the following inequalities a.e. $t \in I$:

$$(D_r\alpha)'(t) \geq f(t, \alpha(t), D_r\alpha(t))$$

and

$$(D_r\beta)'(t) \leq f(t, \beta(t), D_r\beta(t)).$$

For further works in this direction see [62–66].

As we have noted at the beginning of this section, the historical notes concerning this theory have been extracted from [7].

Example 16. Consider the periodic problem (4.32) on the interval $[0, 1]$ for the particular case of

$$f(t, x, y) = (t + x)^3 - \phi_{5/2}(xy),$$

being, for $p > 1$ and $z \in \mathbb{R}$,

$$\phi_p(z) = \begin{cases} |z|^{p-2} z, & z \in \mathbb{R} \setminus \{0\}, \\ 0, & z = 0. \end{cases}$$

It is immediate to verify that all the assumptions of Theorem 69 are fulfilled by taking as lower solution $\alpha = -1$ and as upper solution $\beta = 0$.

So we can ensure the existence of solutions of this problem on the sector $[-1, 0]$.

4.3.2 Existence of Extremal Solutions

M. Cherpion, C. De Coster, and P. Habets proved in [26, Theorem 5] the existence of extremal solutions for the Dirichlet problem. In fact they considered a more general problem: the ϕ-Laplacian equation. In this case they defined a concept of lower and upper solutions in which some kind of angles are allowed. Such definitions were generalized by A. Cabada and R.L. Pouso in [23] to the ϕ-Laplacian equation with nonlinear functional boundary conditions. The definitions are the following ones.

Definition 22. A function $\alpha \in C(I)$ is a lower solution of the Dirichlet problem (4.28) (with $A = B = 0$), if $\alpha(0) \leq 0$, $\alpha(T) \leq 0$ and for any $t_0 \in (0, T)$, either $D_-\alpha(t_0) < D^+\alpha(t_0)$, or there exists an open interval $I_0 \subset I$ such that $t_0 \in I_0$, $\alpha \in W^{2,1}(I_0)$ and, a.e. $t \in I_0$,

$$\alpha''(t) \geq f(t, \alpha(t), \alpha'(t)).$$

Definition 23. A function $\beta \in C(I)$ is a lower solution of the Dirichlet problem (4.28) (with $A = B = 0$), if $\beta(0) \geq 0$, $\beta(T) \geq 0$ and for any $t_0 \in (0, T)$, either $D^-\beta(t_0) > D_+\beta(t_0)$, or there exists an open interval $I_0 \subset I$ such that $t_0 \in I_0$, $\beta \in W^{2,1}(I_0)$ and, a.e. $t \in I_0$,

$$\beta''(t) \leq f(t, \beta(t), \beta'(t)).$$

Here D^+, D_+, D^-, and D_- denote the usual Dini derivatives.

By means of a sophisticated argument, the authors constructed a sequence of upper solutions that converges uniformly to the function defined at each point $t \in I$ as the minimum value attained by all the solutions of problem (4.28) in $[\alpha, \beta]$ at this point. Passing to the limit, they conclude that such function is a solution too. The construction of these upper solutions is valid only in the case that corners are allowed in the definition. The same idea is valid to get a maximal solution.

Similar results, for the periodic boundary conditions, were deduced in [34, Theorem 5.6]. In this case the arguments follow from the finite intersection property of the set of solutions (see [54]).

Remark 48. We point out that [34, Theorem 5.6] can be applied to the problem considered in Example 16. Thus, the existence of extremal solutions of such problem lying on the sector $[-1, 0]$ is deduced.

4.3.2.1 Periodic Boundary Value Problem

Although the arguments developed in [34] can be applied to a problem with nonlinear part depending on the first derivative of the solution, we will show a shorter proof of the existence of extremal solutions without assuming dependence on u'. So, we consider the second order periodic boundary value problem

$$u''(t) = f(t, u(t)), \quad t \in I, \quad u(0) = u(T), \quad u'(0) = u'(T), \qquad (4.47)$$

with f a continuous function.

In this case, following [34], we define a lower and an upper $\mathcal{C}^2$-solution of problem (4.47) as follows.

Definition 24. A function $\alpha \in \mathcal{C}(I)$ such that $\alpha(0) = \alpha(T)$ is a $\mathcal{C}^2$-lower solution of problem (4.47) if its periodic extension on $\mathbb{R}$, defined by $\alpha(t) = \alpha(t + T)$, is such that for any $t_0 \in \mathbb{R}$ one of the two following conditions holds:
1. $D_-\alpha(t_0) < D^+\alpha(t_0)$.
2. There exist an open interval I_0 with $t_0 \in I_0$ and a function $\alpha_0 \in \mathcal{C}^1(I_0, \mathbb{R})$ such that:
   (i) $\alpha(t_0) = \alpha_0(t_0)$ and $\alpha(t) \geq \alpha_0(t)$ for all $t \in I_0$;
   (ii) $\alpha_0''(t_0)$ exists and $\alpha_0''(t_0) \geq f(t_0, \alpha_0(t_0))$.

Definition 25. A function $\beta \in \mathcal{C}(I)$ such that $\beta(0) = \beta(T)$ is a $\mathcal{C}^2$-upper solution of problem (4.47) if its periodic extension on $\mathbb{R}$ is such that for any $t_0 \in \mathbb{R}$ one of the two following conditions holds:

1. $D^-\beta(t_0) > D_+\beta(t_0)$.
2. There exist an open interval I_0 with $t_0 \in I_0$ and a function $\beta_0 \in C^1(I_0, \mathbb{R})$ such that:
    (i) $\beta(t_0) = \beta_0(t_0)$ and $\beta(t) \leq \beta_0(t)$ for all $t \in I_0$;
    (ii) $\beta_0''(t_0)$ exists and $\beta_0''(t_0) \leq f(t_0, \beta_0(t_0))$.

The following result concerning lower and upper C^2-solutions holds.

Lemma 64. *[34, Propositions 2.1 and 2.2]. Let α_1 and α_2 be two C^2-lower solutions. Then*

$$\alpha(t) = \alpha_1 \vee \alpha_2(t) := \max\{\alpha_1(t), \ \alpha_2(t)\} \quad \text{for all } t \in I,$$

is a C^2-lower solution.

Let β_1 and β_2 be two C^2-upper solutions. Then

$$\beta(t) = \beta_1 \wedge \beta_2(t) = \min\{\beta_1(t), \ \beta_2(t)\} \quad \text{for all } t \in I,$$

is a C^2-upper solution.

As we have mentioned before, P. Habets and C. De Coster proved the existence of extremal solutions of (4.47) between α and β in [34, Theorem 2.4] by using the finite intersection property. In [29, Theorem 3.3] J.A. Cid gave a simpler and shorter proof based on Theorem 50. The result is the following one.

Theorem 70. *Let α and β be a pair of C^2-lower and upper solutions of (4.47), such that $\alpha \leq \beta$. Define $E = \{(t, u) \in I \times \mathbb{R} : \alpha(t) \leq u \leq \beta(t)\}$ and assume that f is continuous on E.*

Then the solution set

$$S = \{u \in C^2(I) : \alpha \leq u \leq \beta, \ u \text{ is a solution of } (4.47)\},$$

is a nonempty compact subset of $C(I)$. Moreover, there exist the maximal, u_{max}, and the minimal, u_{min}, solutions of problem (4.6) between α and β, that is, if $u \in S$ then

$$u_{min}(t) \leq u(t) \leq u_{max}(t) \quad \text{for all } t \in I.$$

Proof. It is immediate to verify that S coincides with the set of fixed points P of operator $\mathcal{T} : C(I) \to C(I)$ defined, for each $u \in C(I)$, as

$$\mathcal{T}u(t) = \int_0^T G_P[-1](t, s) \left(f(t, p(s, u(s))) - p(s, u(s))\right) ds,$$

where $G_P[-1](t, s)$ is the Green's function related to the periodic problem

$$u''(t) - u(t) = 0, \quad t \in I, \quad u(0) = u(T), \quad u'(0) = u'(T),$$

and p is the truncated function introduced in (4.37).

Clearly, operator $\mathcal{T}$ is bounded and, in an analogous way to Lemma 63, one can verify that $\mathcal{T}$ is completely continuous. So, we deduce, by Theorem 50, that $P = S$ is a nonempty compact subset of $C(I)$.

Now, we are going to prove that $S = P$ is upward directed with respect to the order induced in $C(I)$ by the cone of non-negative functions. Indeed, given u_1, $u_2 \in S$, Lemma 64 ensures that $\alpha_1 := u_1 \vee u_2 \leq \beta$ is a C^2-lower solution of (4.47) and then, repeating the above argument, there exists a solution u_3 of (4.47) between α_1 and β. Then, since $u_3 \in S$, $u_1 \leq u_3$, and $u_2 \leq u_3$, it follows that $S = P$ is upward directed.

Now, from Theorem 50 (i), we deduce the existence of a greatest fixed point u_{max} of $\mathcal{T}$, which is the maximal solution of (4.47) in the sector enclosed by α and β.

By using a dual argument we prove that problem (4.47) has the minimal solution between α and β. □

4.3.3 Non-Well-Ordered Lower and Upper Solutions

On the contrary to the well ordered case, if α and β are given in the reversed order, i.e. $\alpha \geq \beta$ in I, the existence of solutions of the problem (4.32) cannot be ensured in general. Indeed, let's see the following problem:

$$u''(t) = -u(t) + \sin t, \ t \in [0, 2\pi], \quad u(0) = u(2\pi), \ u'(0) = u'(2\pi).$$

In this case, $\alpha \equiv 1$ and $\beta \equiv -1$ are a pair of reversed ordered lower and upper solutions of this problem. However, if we suppose that such problem has a solution u, then, multiplying both sides of the equation by $\sin t$ and using integration by parts, we arrive at the following contradiction:

$$
\begin{aligned}
\pi = \int_0^{2\pi} \sin^2 t \, dt &= \int_0^{2\pi} u(t) \sin t \, dt + \int_0^{2\pi} u''(t) \sin t \, dt \\
&= \int_0^{2\pi} u(t) \sin t \, dt - \int_0^{2\pi} u(t) \sin t \, dt = 0.
\end{aligned}
$$

This is mainly due to the fact that, as we have seen in the previous chapter, there is no possibility of defining a Dirichlet problem with related positive Green's function. So, the arguments developed in Lemma 60 are not valid for this new situation.

Despite this "general negative existence result", by assuming some additional conditions on the nonlinear part of the equation, it is possible to deduce some existence results when the lower and the upper solution are not well ordered. We will present here an existence result for the homogeneous Dirichlet problem

$$u''(t) = f(t, u(t), u'(t)), \quad t \in I, \quad u(0) = u(T) = 0, \tag{4.48}$$

for $f : I \times \mathbb{R}^2 \to \mathbb{R}$ a continuous function.

The used arguments follow the ones given in [14] for a fourth order problem with clamped beam conditions. Such results can be applied and improved to other situations as the periodic boundary conditions, see the monograph [34, Chapter III]. The first time these arguments were applied is, as far as we know, in the paper [35] of C. De Coster and M. Henrard and they were applied to partial differential equations.

Theorem 71. *Suppose that $f : I \times \mathbb{R} \to \mathbb{R}$ is a continuous function such that for some $A > 0$ we have that*

$$|f(t, x, y)| \le A \quad \text{for all } (t, x, y) \in I \times \mathbb{R}^2. \tag{4.49}$$

Then the following results hold:

(i) *(A priori bounds and existence of solutions) Problem (4.48) has a solution and moreover there exists $k > 0$ such that any solution u of problem (4.48) satisfies $\max\{\|u\|_\infty, \|u'\|_\infty\} \le k$.*

(ii) *(Location of solutions)*

1. *If α is a lower solution, in the sense of Definition 22, then problem (4.48) has a solution u satisfying*

$$\alpha(t) \le u(t) \quad \text{for all } t \in I.$$

2. *If β is an upper solution, in the sense of Definition 23, then problem (4.48) has a solution u satisfying*

$$u(t) \le \beta(t) \quad \text{for all } t \in I.$$

(iii) *(Multiplicity of solutions) Assume now that $f(t, u, v) \equiv f(t, u)$ and there is $M < 0$ for which*

$$f(t, x) - f(t, y) \ge M(y - x) \quad \text{for all } t \in I \text{ and } x \le y.$$

Let α and β be lower and upper solutions such that $\alpha \not\leq \beta$ on I and there are $\bar{t}, \tilde{t} \in (0, T)$ such that

$$\alpha''(\bar{t}) - f(\bar{t}, \alpha(\bar{t})) > 0 > \beta''(\tilde{t}) - f(\tilde{t}, \beta(\tilde{t})). \qquad (4.50)$$

Then problem (4.48) has at least three different solutions u_1, u_2, and u_3, with

$$\alpha(t) \leq u_1(t) \quad and \quad u_2(t) \leq \beta(t) \quad for\ all\ t \in I,$$

and $u_3 \in \tilde{S}$, where

$$\tilde{S} = \{u \in C(I) : \exists\, t_1, t_2 \in I, u(t_1) \geq \beta(t_1),\ \alpha(t_2) \geq u(t_2)\}.$$

Proof. (i) Define the operator $\mathcal{T} : C^1(I) \to C^1(I)$ given for each $h \in C^1(I)$ by

$$\mathcal{T}h(t) = \int_0^T G_D[0](t,s) f(s, h(s), h'(s))\, ds \quad for\ all\ t \in I,$$

where $G_D[0]$ follows the expression (4.46) and is the Green's function related to the Dirichlet problem

$$u''(t) = 0, \quad t \in I, \quad u(0) = u(T) = 0.$$

Analogously to the proof of Lemma 63, one can verify that operator $\mathcal{T}$ is completely continuous and the fixed points of $\mathcal{T}$ are the solutions of (4.48). On the other hand, since f is bounded, there exists $k > 0$ such that

$$\max\{\|\mathcal{T}h\|_\infty, \|(\mathcal{T}h)'\|_\infty\} < k \quad for\ all\ h \in C^1(I).$$

Then we have the a priori bound on the solutions and moreover Schauder's fixed point Theorem (Theorem 33) yields the existence of a solution of problem (4.48).

(ii) We only write the proof for the first case because the second one is similar.
Step 1.- Construction of the modified problem.
    We fix $0 < \varepsilon < (\pi/T)^2$ and for each $r > 0$ we define

$$f_r(t, u, v) = \begin{cases} f(t, -r, v) + \varepsilon (u + r), & if\ u < -r, \\ f(t, u, v), & if\ |u| \leq r, \\ f(t, r, v) + \varepsilon (u - r), & if\ u > r, \end{cases}$$

and consider the modified problem

$$u''(t) = f_r(t, u(t), u'(t)) \quad \text{for all } t \in I, \quad u(0) = u(T) = 0. \tag{4.51}$$

Step 2.- There exists $d > 0$ such that any solution u of (4.51) satisfies $\max\{\|u\|_\infty, \|u'\|_\infty\} \leq d$, *independently of r.*

Applying the classical Wirtinger-type inequality for C^1 functions u defined in I, such that $u(0) = u(T) = 0$ (see inequality (2.1) in [76, Chapter II]. See also the more general (A.2)), we have

$$\left(\frac{\pi}{T}\right)^2 \int_0^T u^2(s)\, ds \leq \int_0^T (u')^2(s)\, ds. \tag{4.52}$$

On the other hand, as f is continuous and bounded, we have that

$$|f_r(t, u, v)| \leq \varepsilon |u| + A \quad \text{for all } (t, u, v) \in I \times \mathbb{R}^2,$$

with A given in (4.49).

Thus, if u is a solution of (4.51), multiplying the equation by u and integrating on I, we have

$$\int_0^T (u')^2(s)\, ds \leq \varepsilon \int_0^T u^2(s)\, ds + A \int_0^T |u(s)|\, ds. \tag{4.53}$$

Now, Hölder's inequality implies that

$$\left(\int_0^T |u(s)|\, ds\right)^2 \leq T \int_0^T u^2(s)\, ds. \tag{4.54}$$

Then, since $0 < \varepsilon < (\pi/T)^2$, from (4.52), (4.53), and (4.54) it follows that u and u' are bounded in $L^2(I)$, independently of r.

Now since $u(0) = 0$, for all $t \in I$ we have that

$$|u(t)| = \left|\int_0^t u'(s)\, ds\right| \leq \int_0^T |u'(s)|\, ds \leq \left(\int_0^T (u')^2(s)\, ds\right)^{\frac{1}{2}} \sqrt{T},$$

which implies that u is also bounded in $L^\infty(I)$, independently of r.

The fact that $u(0) = u(T) = 0$, implies that there is $t_0 \in I$ such that $u'(t_0) = 0$. So

$$|u'(t)| = |u'(t) - u'(t_0)| = \left|\int_{t_0}^t u''(s)\, ds\right| \leq \int_0^T |u''(s)|\, ds$$

$$\leq \int_0^T |f_r(s, u(s), u'(s))|\, ds \leq \varepsilon \int_0^T |u(s)|\, ds + A\, T,$$

which implies, using (4.54) and the boundedness of u in $L^2(I)$ (without dependence on r), that u' is also bounded in $L^\infty(I)$, independently of r.

As a consequence, we have that there exists $d > 0$, that does not depend on r, such that

$$\max\{\|u\|_\infty, \|u'\|_\infty\} \leq d.$$

Step 3.- There exists a solution u of (4.51) for all r large enough.

Let $r > \|\alpha\|_\infty$. It is immediate to verify that α remains as a lower solution of problem (4.51) and

$$\beta_1(t) = \frac{A}{\varepsilon} + r + 1 > r > \alpha(t), \quad t \in I,$$

is an upper solution of problem (4.51).

Now, using [26, Theorem 5], with f_r instead of f, it is ensured the existence of a solution of the modified problem (4.51) in $[\alpha, \beta_1]$.

Conclusion: By steps 1 and 2 taking $r > d$ we obtain the existence of a solution $u \geq \alpha$ of (4.51) with $\max\{\|u\|_\infty, \|u'\|_\infty\} \leq d < r$. Hence u is also a solution of the original problem (4.48) and assertion (*ii*) is proved.

(*iii*) To prove the third case, we use the Amann's "three fixed points" Theorem 44.

The existence of two solutions follows from (*ii*) and the fact that $\alpha \not\leq \beta$ on I.

Choose $r > \max\{d, \|\alpha\|_\infty, \|\beta\|_\infty\}$, $0 < \varepsilon < -M$, and define

$$\alpha_1(t) = -\frac{A}{\varepsilon} - r - 1 \quad \text{and} \quad \beta_1(t) = \frac{A}{\varepsilon} + r + 1,$$

which are clearly lower and upper solutions, respectively, for the modified problem (4.51).

Let $\mathcal{T}_M : C^1(I) \to C^1(I)$, be defined as

$$\mathcal{T}_M h(t) = \int_0^T G_D[M](t, s)\, \big(f_r(s, h(s)) + M\, h(s)\big)\, ds \quad \text{for all } t \in I,$$

where $G_D[M]$ is the Green's function related to the Dirichlet problem

$$u''(t) + M u(t) = \sigma(t), \quad t \in I, \quad u(0) = u(T) = 0, \qquad (4.55)$$

which (see [12]) is given by the following expression (denoting $M = -m^2$)

$$G_D[M](t, s) = -\frac{2\,e^{m\,T}}{m\,(e^{2\,m\,T} - 1)} \begin{cases} \sinh(m\,s)\sinh(m(T-t)), & 0 \le s \le t \le T, \\ \sinh(m\,t)\sinh(m(T-s)), & 0 < t < s \le T. \end{cases}$$

It is obvious that $G_D[M](t, s) < 0$ for all $(t, s) \in (0, T) \times (0, T)$ and, moreover

$$\frac{\partial}{\partial t}G_D[M](t, s)|_{t=0} < 0 < \frac{\partial}{\partial t}G_D[M](t, s)|_{t=T}, \quad \text{for all } s \in (0, T).$$

As a direct consequence, it is not difficult to verify that if $\sigma \ge 0$, $\sigma \not\equiv 0$ on $(0, T)$, then the unique solution u of problem (4.55) satisfies that $u < 0$ on $(0, T)$ and $u'(0) < 0 < u'(T)$. In case of $u(0) < 0$, we can ensure that $u < 0$ on $[0, T)$, but nothing can be ensured for the value of $u'(0)$. A similar comment is valid if $u(T) < 0$.

We have several possibilities:

1. If $\alpha(0) < 0$, $\alpha(T) < 0$, $\beta(0) > 0$ and $\beta(T) > 0$ we work in the space $\mathcal{C}(I)$ and set

$$X = \{u \in \mathcal{C}(I) : \alpha_1(t) \le u(t) \le \beta_1(t) \text{ in } I\},$$
$$X_1 = \{u \in X : \alpha(t) \le u(t) \le \beta_1(t) \text{ in } I\},$$
$$X_2 = \{u \in X : \alpha_1(t) \le u(t) \le \beta(t) \text{ in } I\},$$
$$O_1 = \{u \in X : \alpha(t) < u(t) < \beta_1(t) \text{ in } I\},$$
$$O_2 = \{u \in X : \alpha_1(t) < u(t) < \beta(t) \text{ in } I\}.$$

2. If (to fix ideas) $\alpha(0) = \alpha(T) = 0$, $\beta(0) > 0$ and $\beta(T) > 0$ (the remaining cases being treated with the obvious changes), then we work on the space

$$\mathcal{C}_0^1(I) := \{u \in \mathcal{C}^1(I) : u(0) = u(T) = 0\},$$

define

$$D = \max\left\{d, \sup_{\alpha_1 \le h \le \beta_1} \|\mathcal{T}_M h\|_{\mathcal{C}^1(I)}\right\}$$

and consider the sets

$$X = \{u \in \mathcal{C}_0^1(I) : \alpha_1(t) \le u(t) \le \beta_1(t) \text{ in } I, \ \|u'\|_\infty \le D\},$$
$$X_1 = \{u \in X : \alpha(t) \le u(t) \le \beta_1(t) \text{ in } I\},$$

$$X_2 = \{u \in X : \alpha_1(t) \le u(t) \le \beta(t) \text{ in } I\},$$
$$O_1 = \{u \in X : \alpha(t) < u(t) < \beta_1(t) \text{ in } (0, T), \ u'(0) > \alpha'(0), \ u'(T) < \alpha'(T)\},$$
$$O_2 = \{u \in X : \alpha_1(t) < u(t) < \beta(t) \text{ in } (0, T)\}.$$

In any case X_i is a closed, bounded, convex subset of X and O_i is an open set with $O_i \subset X_i$.

Let's see that $\mathcal{T}_M(X_i) \subset X_i$, $i = 1, 2$. Indeed, from the monotone non-increasing character of $f(t, x) + Mx$, the fact that $\varepsilon + M < 0$, and the negativeness of function $G_D[M]$, it is immediate to verify that operator $\mathcal{T}_M$ is monotone non-decreasing on X.

Consider the case 2. and $i = 1$ (the other ones are analogous), by definition, we have that $(\mathcal{T}_M \alpha)(0) = (\mathcal{T}_M \alpha)(T) = 0$ and, moreover, for all $t \in (0, T)$,

$$
\begin{aligned}
(\mathcal{T}_M \alpha)''(t) + M \mathcal{T}_M \alpha(t) &= f_r(t, \alpha(t)) + M\alpha(t) \\
&= f(t, \alpha(t)) + M\alpha(t) \le \alpha''(t) + M\alpha(t),
\end{aligned}
$$

being the last inequality strict at some subinterval of I.

Therefore, as we have noticed before, it is fulfilled that

$$\mathcal{T}_M \alpha > \alpha \text{ on } (0, T), \ (\mathcal{T}_M \alpha)'(0) > \alpha'(0) \text{ and } (\mathcal{T}_M \alpha)'(T) < \alpha'(T).$$

Analogously, one can verify that

$$\mathcal{T}_M \beta_1 < \beta_1 \text{ on } I.$$

So, the monotonicity of operator $\mathcal{T}_M$ implies that $\mathcal{T}_M(X_1) \subset X_1$.

We claim that $\mathcal{T}_M$ has no fixed point in $X_i \setminus O_i$, $i = 1, 2$.

Indeed, if $u \in X_1$ is a fixed point of $\mathcal{T}_M$ then the function $z = \alpha - u$ satisfies the inequality $z'' + Mz \ge 0$ in I, with the inequality strict at some subinterval of I, and we conclude that $z < 0$ on $(0, T)$. Moreover, either $z(0) < 0$ and $z(T) < 0$ (Case 1) or $z(0) = z(T) = 0$ and $z'(0) < 0 < z'(T)$ (Case 2).

Similarly, one can verify that $u < \beta_1$ on I and so $u \in O_1$.

The same is true for X_2 and O_2.

Therefore Theorem 44 implies the existence of three solutions with the desired properties. $\qquad \square$

As it has been remarked before, previous result can be applied to a wider set of boundary conditions and to not necessarily continuous functions. One of the main representative of such results, for the periodic conditions, can be found in [34]. In such a case C. De Coster and P. Habets introduced the following concept of lower and upper solutions.

Definition 26. A function $\alpha \in C(I)$, such that $\alpha(0) = \alpha(T)$, is a lower solution of the periodic problem (4.32), if its periodic extension to $\mathbb{R}$, defined by $\alpha(t) = \alpha(t + T)$, is such that and for any $t_0 \in \mathbb{R}$, either $D_-\alpha(t_0) < D^+\alpha(t_0)$, or there exists an open interval $I_0 \subset I$ such that $t_0 \in I_0$, $\alpha \in W^{2,1}(I_0)$ and, a.e. $t \in I_0$,

$$\alpha''(t) \geq f(t, \alpha(t), \alpha'(t)).$$

Definition 27. A function $\beta \in C(I)$ such that $\beta(0) = \beta(T)$, is a lower solution of the periodic problem (4.32), if its periodic extension to $\mathbb{R}$, defined by $\beta(t) = \beta(t + T)$, and for any $t_0 \in (0, T)$, either $D^-\beta(t_0) > D_+\beta(t_0)$, or there exists an open interval $I_0 \subset I$ such that $t_0 \in I_0$, $\beta \in W^{2,1}(I_0)$ and, a.e. $t \in I_0$,

$$\beta''(t) \leq f(t, \beta(t), \beta'(t)).$$

The existence result is the following

Theorem 72. *Let α and β be a pair of lower and upper solutions of the periodic problem (4.32) such that $\alpha \not\leq \beta$ on I. Define $A \subset I$ (respectively $B \subset I$) to be the set of points where α (respectively β) is derivable.*

Assume that f is a Carathéodory function and there exists $N \in L^1(I)$, $N > 0$ a.e. I such that for a.e. $t \in A$ (respectively a.e. $t \in B$)

$$f(t, \alpha(t), \alpha'(t)) \geq -N(t) \quad (\text{respectively } f(t, \beta(t), \beta'(t)) \leq N(t)).$$

Assume further that for some $h \in L^1(I)$ either

$$f(t, u, v) \leq h(t) \quad \text{on } I \times \mathbb{R}^2,$$

or

$$f(t, u, v) \geq h(t) \quad \text{on } I \times \mathbb{R}^2.$$

Then there exists a solution of (4.32) in

$$\mathcal{S} := \{u \in C^1(I), \text{ such that there exist } t_1, \ t_2 \in I, \text{ for which } u(t_1) \geq \beta(t_1),$$
$$u(t_2) \leq \alpha(t_2)\}.$$

Next example is an adaptation of [14, Example 3.1].

Example 17. Consider the function $f : I \times \mathbb{R} \to \mathbb{R}$ defined by

$$f(t, x) = (-4\pi^2 + 2\sin(2\pi t))\,\delta(x) + 2\cos^2(2\pi t) - 2 + g(x),$$

where $\delta : \mathbb{R} \to \mathbb{R}$ is defined as

$$\delta(x) = \begin{cases} -1, & \text{if } x < -1, \\ x, & \text{if } |x| \leq 1, \\ 1, & \text{if } x > 1, \end{cases}$$

and $g : \mathbb{R} \to \mathbb{R}$ is a continuous and bounded function that satisfies that $g(0) = 0$ and $0 < g(x) < 1$ for all $x \neq 0$, $|x| < 1$, and for some $K < 0$ the function $g(x) + Kx$ is monotone non-increasing in x.

It is easy to check that $\alpha(t) = \sin(2\pi t)$ and $\beta(t) = 0$ for all $t \in [0, 1]$ are lower and upper solutions, respectively, that satisfy (4.50), and $\alpha \not\leq \beta$. Moreover f satisfies the conditions of Theorem 71, (*iii*) for $M = K$. Therefore it can be ensured the existence of at least three solutions for the considered problem.

4.4 MONOTONE ITERATIVE TECHNIQUES

As we have seen in the previous section, the existence of a pair of well ordered lower and upper solutions is a fundamental tool to deduce the existence of solutions of second order boundary value problems lying between both functions. Moreover, it is possible to ensure that there are extremal solutions lying in the sector formed by the lower solution α and the upper solution β. At this stage, we think on the possibility of obtaining the expression of the extremal solutions as a uniform limit of a sequence of solutions of related linear problems. To this end, following the ideas developed in [20], we can present a boundary value problem of the type (4.28) (either (4.31) or (4.32)), in a general framework, as an abstract equation

$$Lu = Nu, \tag{4.56}$$

where $L : D(L) \subset H \to H$ is a linear operator, $N : D(N) \subset H \to H$ is a nonlinear operator, and H is a given Hilbert space with functions defined on Ω, a bounded and open set in $\mathbb{R}^m$.

If there exists the inverse of L, then this abstract problem is equivalent to

$$u = L^{-1}(Nu),$$

and we have to find a fixed point for the operator $L^{-1} \circ N$.

However, in many practical situations, we have that L is not invertible but there exists $(L + \lambda I)^{-1}$ for some $\lambda \in \mathbb{R}$. This is the case, for instance, of operator $Lu = u''$ on the domain $D(L) = \{u \in W^{2,2}(I), u(0) = u(T), u'(0) = u'(T)\}$, that has been treated in previous section.

In this situation, Eq. (4.56) is equivalent to

$$u = (L + \lambda I)^{-1}(N + \lambda I) u \equiv A_\lambda u. \tag{4.57}$$

So, in order to approximate the extremal fixed points of operator A_λ, we apply the classical monotone iterative method developed by Amman in Theorem 51.

4.4.1 Well Ordered Lower and Upper Solutions

It is clear that to ensure the monotone non-decreasing character of operator A_λ, defined in (4.57), we can impose the two following conditions:

(a) The linear operator $L + \lambda I$ is inverse negative on $D(L)$, that is,

$$u \in D(L), \ Lu + \lambda u \geq 0 \text{ on } \Omega \text{ implies that } u \leq 0 \text{ on } \Omega. \tag{4.58}$$

This means that the Green's function (in case it exists) is nonpositive (see Chapter 3 and [6,8]).

(b) The nonlinear operator $N + \lambda I$ is monotone non-increasing on $D(N)$.

To fix ideas, consider the Hilbert space $H = L^2(\Omega)$ with the usual inner product and norm. We write $u \geq 0$ on Ω if $u(x) \geq 0$ for a.e. $x \in \Omega$.

In many cases the operator $(L + \lambda I)^{-1}$ is a Hilbert-Schmidt operator. This is the case for many boundary value problems for ordinary and integro-differential equations and for the Dirichlet problem for elliptic partial differential equations with $m \leq 3$. In this situation, there exists a function $G[\lambda] \in L^2(\Omega \times \Omega)$ such that the operator $G_\lambda = (L + \lambda I)^{-1}$ is precisely the integral operator with kernel $G[\lambda]$, that is, for $\sigma \in H$,

$$[G_\lambda \sigma](x) = \int_\Omega G[\lambda](x, y) \sigma(y) \, dy. \tag{4.59}$$

So, one can define $\alpha \in D(L) \cap D(N)$ as a lower solution for (4.56) if

$$L\alpha \geq N\alpha \text{ on } \Omega. \tag{4.60}$$

Analogously, $\beta \in D(L) \cap D(N)$ is an upper solution for (4.56) whenever

$$L\beta \leq N\beta \text{ on } \Omega. \tag{4.61}$$

Note that (4.60) implies that

$$L\alpha + \lambda\alpha \geq N\alpha + \lambda\alpha.$$

Hence, there exists $\xi \in H$, $\xi \geq 0$ on Ω, such that

$$L\alpha + \lambda\alpha = N\alpha + \lambda\alpha + \xi,$$

and

$$\alpha = G_\lambda[N\alpha + \lambda\alpha + \xi] = A_\lambda\alpha + G_\lambda\xi.$$

If condition (a) holds, which is equivalent to the fact that $G[\lambda] \leq 0$ on $\Omega \times \Omega$, we have that

$$L\alpha \geq N\alpha \text{ on } \Omega \text{ implies that } \alpha \leq A_\lambda\alpha \text{ on } \Omega.$$

Analogously, we can deduce that

$$L\beta \leq N\beta \text{ on } \Omega \text{ implies that } \beta \geq A_\lambda\beta \text{ on } \Omega.$$

Thus, assuming that the lower and upper solutions are well ordered, i.e., $\alpha \leq \beta$ on Ω, we conclude that if condition (b) is fulfilled, we can ensure that A_λ is monotone non-decreasing and $A_\lambda[\alpha, \beta] \subset [\alpha, \beta]$, where

$$[\alpha, \beta] = \{v \in H : \alpha \leq v \leq \beta \text{ on } \Omega\}.$$

As direct consequence of Theorem 51, we arrive at the following abstract result.

Theorem 73. *Consider the nonlinear equation (4.56) and suppose that α, $\beta \in D(L) \cap D(N)$, $\alpha \leq \beta$ on Ω, are lower and upper solutions of (4.56), in the sense of definitions (4.60) and (4.61) respectively.*

Assume that there exists $\lambda \in \mathbb{R}$ such that the maximum principle (4.58) is valid, G_λ is a Hilbert-Schmidt operator given by (4.59), $[\alpha, \beta] \subset D(N)$ and $N : [\alpha, \beta] \to H$ is continuous.

Moreover, suppose that

$$Nu - Nv \leq -\lambda(u - v)$$

for any $u, v \in H$ with $\alpha \leq v \leq u \leq \beta$ on Ω.

Then, there exist two monotone sequences $\{\alpha_n\}$ and $\{\beta_n\}$, which converge in H to ϕ and ψ respectively, with $\alpha_0 = \alpha$ and $\beta_0 = \beta$. Here ϕ and ψ are the minimal and maximal solutions of (4.56), respectively, between α and β, that is, if u is a solution of (4.56) with $\alpha \leq u \leq \beta$ on Ω, then $\phi \leq u \leq \psi$ on Ω.

Under this framework, we can consider all the linear boundary conditions studied in Chapter 3 for the Hill's equation: periodic, Neumann, Dirichlet, and mixed.

Taking, for instance, the periodic conditions (the other ones can be treated in the same manner), we can study the problem

$$u''(t) + a(t)\, u(t) = f(t, u(t)), \text{ a.e. } t \in I, \quad u(0) = u(T), \; u'(0) = u'(T).$$
$$(4.62)$$

In this case, we assume $a \in L^2(I)$ and f a L^2-Carathéodory function, i.e., it satisfies conditions in Definition 11, with $h_R \in L^2(I)$ in assumption (iii).

The involved operators and domains are the following ones:

$$L\, u(t) := u''(t) + a(t)\, u(t) \text{ defined on}$$
$$D(L) := \{u \in W^{2,2}(I), \; u(0) = u(T), \; u'(0) = u'(T)\}$$

and

$$N\, u(t) := f(t, u(t)) \text{ defined on } D(N) := C(I).$$

By Lemma 25 we have that condition (4.58) is fulfilled if and only if $\lambda < \lambda_0^P[a]$, with $\lambda_0^P[a]$ the smallest eigenvalue of the periodic problem

$$u''(t) + (a(t) + \lambda)\, u(t) = 0, \quad \text{a.e. } t \in I, \quad u(0) = u(T), \; u'(T) = u'(T).$$

As a consequence, if there exist α, $\beta \in W^{2,2}(I)$, $\alpha \leq \beta$ in I, satisfying:

$$\alpha''(t) + a(t)\, \alpha(t) \geq f(t, \alpha(t)), \text{ a.e. } t \in I, \quad \alpha(0) = \alpha(T), \; \alpha'(0) = \alpha'(T)$$

and

$$\beta''(t) + a(t)\, \beta(t) \leq f(t, \beta(t)), \text{ a.e. } t \in I, \quad \beta(0) = \beta(T), \; \beta'(0) = \beta'(T),$$

and, moreover, function f satisfies

$$f(t, u) - f(t, v) \leq -\lambda\, (u - v), \text{ a.e. } t \in I, \; \alpha(t) \leq v \leq u \leq \beta(t), \qquad (4.63)$$

for some $\lambda < \lambda_0^P[a]$, we have that there exist two monotone sequences $\{\alpha_n\}$ and $\{\beta_n\}$, which converge in $L^2(I)$ to ϕ and ψ respectively, with $\alpha_0 = \alpha$ and $\beta_0 = \beta$. Here ϕ and ψ are the minimal and maximal solutions of (4.62), respectively, lying between α and β.

Remark 49. Notice that, if f is differentiable with respect to x, condition (4.63) reads as follows:

$$\frac{\partial f}{\partial x}(t, x) \leq -\lambda \text{ for a.e. } t \in I \text{ and all } \alpha(t) \leq x \leq \beta(t),$$

for some $\lambda < \lambda_0^P[a]$.

It is obvious that, since α and β are continuous on I, such property is always true whenever $f(t, \cdot) \in C^1(\mathbb{R})$ a.e. $t \in I$.

As we have proved in Chapter 3, the Green's function related to Dirichlet, Neumann or mixed problems is negative on $I \times I$ for all λ strictly less than the first eigenvalue of the related linear problem in each case. So, the monotone method and, as a consequence, the existence of extremal solutions of the considered problems can be ensured as for these cases.

Example 18. Consider the periodic problem (4.62) for the particular case of $a \equiv 0$, $T = 1$, and

$$f(t, x) = \phi_q(t) + \phi_p(x), \quad q > 1, \ p \geq 2,$$

with ϕ_m defined in Example 16.

It is immediate to verify that $\alpha = -1$ and $\beta = 0$ are a pair of well ordered lower and upper solutions for this problem.

Moreover $\lambda_0^P[0] = 0$ and

$$\frac{\partial f}{\partial x}(t, x) = (p - 1)|x|^{p-2} \text{ for all } t \in I \text{ and } x \in \mathbb{R}.$$

Thus, since $p \geq 2$, (4.63) holds for all $\lambda \leq 1 - p$, and we can ensure the existence of extremal solutions of this problem on the sector $[-1, 0]$ and that such extremal solutions can be obtained by means of a sequence of functions that solve related linear problems.

However, if $1 < p < 2$, we have that $\frac{\partial f}{\partial x}(t, x)$ is unbounded in $x \in [-1, 0]$ and we cannot ensure the validity of Theorem 73. In this case, we can ensure the existence of extremal solutions of the considered problem from Theorem 70.

4.4.2 Reversed Ordered Lower and Upper Solutions

As we have noticed in the previous subsection, when the lower and the upper solutions are given in the reversed order, i.e. $\alpha \geq \beta$, the existence of solution of the considered problem is not guaranteed in general. Despite this, we can develop the monotone method to this situation and, as a consequence, ensure the existence of extremal solutions on the sector $[\beta, \alpha]$.

In this situation, we must assume the corresponding dual hypotheses, which are sufficient to ensure that operator A_λ is monotone nondecreasing:

(a')   The linear operator $L + \lambda I$ is inverse positive on $D(L)$, that is,

$$u \in D(L), \quad Lu + \lambda u \geq 0 \text{ on } \Omega \text{ implies that } u \geq 0 \text{ on } \Omega. \qquad (4.64)$$

This means that the Green's function (in case it exists) is non-negative (see Chapter 3 [6,8]).

(b')   The nonlinear operator $N + \lambda I$ is monotone non-decreasing on $D(N)$.

Arguing as in the previous subsection, and as a direct consequence of Theorem 51, we arrive at the following abstract result (see [80, Theorem 3.1] and [20, Theorem 2.1]).

Theorem 74. *Consider the nonlinear equation (4.56) and suppose that α, $\beta \in D(L) \cap D(N)$, $\alpha \geq \beta$ on Ω, are lower and upper solutions of (4.56), in the sense of definitions (4.60) and (4.61) respectively.*

Assume that there exists $\lambda \in \mathbb{R}$ such that the anti-maximum principle (4.64) is valid, G_λ is a Hilbert-Schmidt operator given by (4.59), $[\beta, \alpha] \subset D(N)$ and $N : [\beta, \alpha] \to H$ is continuous.

Moreover, suppose that

$$Nu - Nv \geq -\lambda (u - v)$$

for any $u, v \in H$ with $\beta \leq v \leq u \leq \alpha$ on Ω.

Then, there exist two monotone sequences $\{\alpha_n\}$ and $\{\beta_n\}$ which converge in H to ϕ and ψ respectively, with $\alpha_0 = \alpha$ and $\beta_0 = \beta$. Here ϕ and ψ are the maximal and minimal solutions of (4.56), respectively, between β and α, that is, if u is a solution of (4.56) with $\beta \leq u \leq \alpha$ on Ω, then $\phi \leq u \leq \psi$ on Ω.

In this case, as we have seen in Chapter 3, the Green's function related to the Hill's equation with Dirichlet or mixed boundary conditions is never positive on its whole square of definition. Thus, Theorem 74 cannot be applied to these classes of boundary conditions. However, it can be applied to the Neumann and periodic ones. In particular, considering the periodic problem (4.62), we know, by Lemma 25 again, that condition (4.64) is fulfilled if and only if $\lambda_0^P[a] < \lambda \leq \lambda_0^A[a]$, being $\lambda_0^A[a]$ the smallest eigenvalue of the anti-periodic equation

$$u''(t) + (a(t) + \lambda)\,u(t) = 0, \quad \text{a.e. } t \in I, \qquad u(0) = -u(T), \ u'(0) = -u'(T).$$

As a consequence, if there exists a pair of lower and upper solutions defined as in the previous subsection, with $\alpha \geq \beta$ in I, and function f satisfies

$$f(t, u) - f(t, v) \geq -\lambda\,(u - v), \text{ a.e. } t \in I, \quad \beta(t) \leq v \leq u \leq \alpha(t), \qquad (4.65)$$

for some $\lambda_0^P[a] < \lambda \leq \lambda_0^A[a]$, we have that there exist two monotone sequences $\{\alpha_n\}$ and $\{\beta_n\}$, which converge in $L^2(I)$ to ϕ and ψ respectively, with $\alpha_0 = \alpha$ and $\beta_0 = \beta$. Here ϕ and ψ are the maximal and minimal solutions of (4.62), respectively, between β and α.

Remark 50. Notice that, if f is differentiable with respect to x, condition (4.65) reads as follows:

$$\frac{\partial f}{\partial x}(t, x) \geq -\lambda \text{ for a.e. } t \in I \text{ and all } \beta(t) \leq x \leq \alpha(t),$$

for some $\lambda_0^P[a] < \lambda \leq \lambda_0^A[a]$.

Example 19. Consider the periodic problem (4.62) for the particular case of $a \equiv 0$, $T = 1$, and

$$f(t, x) = \phi_q(t) - \phi_p(x), \quad q > 1, \ p \geq 2,$$

with ϕ_m defined in Example 16.

It is immediate to verify that $\alpha = 1$ and $\beta = 0$ are a pair of reversed ordered lower and upper solutions for this problem.

Moreover $\lambda_0^P[0] = 0$, $\lambda_0^A[0] = \pi^2$, and

$$\frac{\partial f}{\partial x}(t, x) = -(p - 1)\,|x|^{p-2} \text{ for all } t \in I \text{ and } x \in \mathbb{R}.$$

Thus, since $p \geq 2$, (4.65) holds for all $\lambda \geq p - 1$.

Thus, provided $2 \leq p \leq \pi^2 + 1$, we can ensure the existence of extremal solutions of this problem on the sector $[0, 1]$ and that such extremal solutions can be obtained by means of a sequence of functions that solve related linear problems.

However, if $1 < p < 2$, since $\frac{\partial f}{\partial x}(t, x)$ is unbounded in $x \in [0, 1]$ we cannot ensure the validity of Theorem 74. Moreover, in this case, Theorem 70 is not valid either, and we cannot ensure the existence of solutions for this problem lying between 0 and 1.

Concerning the Neumann problem, by making the natural changes on the definitions of α and β:

$$\alpha''(t) + a(t)\,\alpha(t) \geq f(t, \alpha(t)), \text{ a.e. } t \in I, \quad \alpha'(0) = \alpha'(T) = 0$$

and

$$\beta''(t) + a(t)\,\beta(t) \leq f(t, \beta(t)), \text{ a.e. } t \in I, \quad \beta'(0) = \beta'(T) = 0,$$

and using Theorem 18, the same existence result is valid for the Neumann problem when $\lambda_0^N[a] < \lambda < \min\left\{\lambda_0^{M_1}[a], \lambda_0^{M_2}[a]\right\}$. Here, $\lambda_0^N[a]$ is the smallest eigenvalue of the Neumann problem

$$u''(t) + (a(t) + \lambda)\,u(t) = 0, \quad \text{a.e. } t \in I, \quad u'(0) = u'(T) = 0,$$

$\lambda_0^{M_1}[a]$ is the smallest eigenvalue of the mixed problem

$$u''(t) + (a(t) + \lambda)\,u(t) = 0, \quad \text{a.e. } t \in I, \quad u'(0) = u(T) = 0$$

and $\lambda_0^{M_2}[a]$ is the smallest eigenvalue of the mixed problem

$$u''(t) + (a(t) + \lambda)\,u(t) = 0, \quad \text{a.e. } t \in I, \quad u(0) = u'(T) = 0.$$

As we have noticed along this subsection, the iteration technique is valid whenever the related Green's function is non-negative on its square of definition for some suitable values of λ for which the nonlinear part of the equation satisfies the monotonicity condition (4.65).

In the sequel, we present an example in which it is showed that if such property does not hold, then the monotone iterative technique can fail. The result, given in [5, Theorem 3.1], is the following one:

Let $m_1^2 > (\pi/T)^2 \equiv \lambda_0^P[0]$, $m_1\,T/(2\pi) \notin \mathbb{N}$, and consider the following periodic problem

$$u''(t) = -m_1^2\,u(t) + \sigma(t) \equiv f(t, u(t)), \; t \in I, \quad u(0) = u(T), \; u'(0) = u'(T),$$

$$(4.66)$$

with σ defined as follows

$$\sigma(s) = \begin{cases} m_1^2 \exp\left(1 - \dfrac{1}{1-\left[\frac{s-s_0}{\epsilon}\right]^2}\right) & \text{if } |s - s_0| < \epsilon, \\ 0 & \text{if } |s - s_0| > \epsilon. \end{cases}$$

Since m_1 is a nonresonant value for problem (4.66), such problem has a unique solution, which is given by the expression

$$u(t) = \int_0^T G_P[m_1^2](t, s)\,\sigma(s)\,ds.$$

From Lemma 25, we know that $G_P[m_1^2]$ changes its sign on $I \times I$ for all $m > m_1$. So, we can choose s_0 and ϵ such that the related Green's function $G_P[m_1^2]$ satisfies $G_P[m_1^2](0, s) < 0$ in $(s_0 - \epsilon, s_0 + \epsilon)$.

Obviously $\alpha \equiv 1$ is a lower solution and $\beta \equiv 0$ is an upper solution for this problem, and f satisfies condition (4.65) for $\lambda \geq m_1^2$. However it is clear that the unique solution satisfies that $u(0) < 0$, i.e., there is no solution of the considered problem on the sector $[\alpha, \beta]$.

The same example, with obvious adaptations, is valid for the Neumann case.

4.4.2.1 Final Remarks

It is important to point out that if we consider other kind of linear boundary conditions for which the Green's function changes its sign on $I \times I$, we can construct similar examples of the optimality of the constant sign of the Green's function. This property is valid for both cases $\alpha \leq \beta$ and $\alpha \geq \beta$.

In [20] it is developed a monotone iterative technique for the case in which the Green's function changes its sign. In this case the definition of the lower and upper solution involve the positive and negative part of the related Green's function and is valid only when the uniqueness of solution is ensured.

Monotone iterative techniques with dependence on u' have been developed for the well ordered case in [60] for continuous functions and in [19, 97] for Carathéodory ones.

In [81] it is proved the validity of the monotone method for well ordered lower and upper solutions by deducing a maximum principle for the nonlinear operator $u'' + k|u'| + mu$. In [16] it is obtained an anti-maximum principle for such operator and, as a consequence, the reversed order case is considered. A generalization of these results has been done in [32].

REFERENCES

[1] H. Amann, On the number of solutions of nonlinear equations in ordered Banach spaces, J. Funct. Anal. 11 (1972) 346–384.

[2] H. Amann, Fixed point equations and nonlinear problems in ordered Banach spaces, SIAM Rev. 18 (4) (1976) 620–709.

[3] V. Bevc, J.L. Palmer, C. Süsskind, On the design of the transition region of axi-symmetric magnetically focusing beam valves, J. Br. Inst. Radio Eng. 18 (1958) 696–708.

[4] S.R. Bernfeld, V. Lakshmikantham, An Introduction to Nonlinear Boundary Value Problems, Math. Sci. Eng., vol. 109, Academic Press, New York, 1974.

[5] A. Cabada, The method of lower and upper solutions for second, third, fourth and higher order boundary value problems, J. Math. Anal. Appl. 185 (1994) 302–320.

[6] A. Cabada, The method of lower and upper solutions for third order periodic boundary value problems, J. Math. Anal. Appl. 195 (1995) 568–589.

[7] A. Cabada, An overview of the lower and upper solutions method with nonlinear boundary value conditions, Bound. Value Probl. (2011) 893753.

[8] A. Cabada, Green's Functions in the Theory of Ordinary Differential Equations, Springer Briefs Math., Springer, New York, 2014.

[9] A. Cabada, J.A. Cid, A note on fixed point theorems for T-monotone operators, Comput. Math. Appl. 47 (2004) 853–857.

[10] A. Cabada, J.A. Cid, Existence of a non-zero fixed point for nondecreasing operators via Krasnosel'skiĭ's fixed point theorem, Nonlinear Anal. 71 (2009) 2114–2118.

[11] A. Cabada, J.A. Cid, Existence and multiplicity of solutions for a periodic Hill's equation with parametric dependence and singularities, Abstr. Appl. Anal. (2011) 545264.

[12] A. Cabada, J.A. Cid, B. Máquez-Villamarín, Computation of Green's functions for boundary value problems with Mathematica, Appl. Math. Comput. 219 (4) (2012) 1919–1936.

[13] A. Cabada, J.A. Cid, G. Infante, New criteria for the existence of non-trivial fixed points in cones, Fixed Point Theory Appl. 2013 (2013) 125, 12 pp.

[14] A. Cabada, J.A. Cid, L. Sanchez, Positivity and lower and upper solutions for fourth order boundary value problems, Nonlinear Anal. 67 (5) (2007) 1599–1612.

[15] A. Cabada, Z. Hamdi, Existence results for nonlinear fractional Dirichlet problems on the right side of the first eigenvalue, Georgian Math. J. 24 (1) (2017) 41–53.

[16] A. Cabada, P. Habets, S. Lois, Monotone method for the Neumann problem with lower and upper solutions in the reverse order, Appl. Math. Comput. 117 (1) (2001) 1–14.

[17] A. Cabada, P. Habets, R.L. Pouso, Optimal existence conditions for ϕ-Laplacian equations with upper and lower solutions in the reversed order, J. Differential Equations 166 (2) (2000) 385–401.

[18] A. Cabada, S. Heikkilä, Extremality results for discontinuous explicit and implicit diffusion problems, J. Comput. Appl. Math. 143 (1) (2002) 69–80.

[19] A. Cabada, Lois, S. An, Existence result for nonlinear second order periodic boundary value problems with lower and upper solutions in reverse order, Eng. Simul. 16 (1999) 319–328.

[20] A. Cabada, J.J. Nieto, Fixed points and approximate solutions for nonlinear operator equations. Fixed point theory with applications in nonlinear analysis, J. Comput. Appl. Math. 113 (1–2) (2000) 17–25.

[21] A. Cabada, D. O'Regan, R.L. Pouso, Second order problems with functional conditions including Sturm-Liouville and multipoint conditions, Math. Nachr. 281 (9) (2008) 1254–1263.

[22] A. Cabada, R.L. Pouso, Existence results for the problem $(\phi(u'))' = f(t, u, u')$ with nonlinear boundary conditions, Nonlinear Anal. 35 (2) (1999) 221–231.

[23] A. Cabada, R.L. Pouso, Extremal solutions of strongly nonlinear discontinuous second-order equations with nonlinear functional boundary conditions, Nonlinear Anal. 42 (8) (2000) 1377–1396.

[24] A. Cabada, L. Saavedra, The eigenvalue characterization for the constant sign Green's functions of $(k, n - k)$ problems, Bound. Value Probl. (2016) 44, 35 pp.

[25] S. Carl, S. Heikkilä, Nonlinear Differential Equations in Ordered Spaces, Chapman & Hall/CRC Monogr. Surv. Pure Appl. Math., vol. 111, Chapman & Hall/CRC, Boca Raton, FL, 2000.

[26] M. Cherpion, C. De Coster, P. Habets, Monotone iterative methods for boundary value problems, Differential Integral Equations 12 (3) (1999) 309–338.

[27] J. Chu, Z. Zhang, Periodic solutions of singular differential equations with sign-changing potential, Bull. Aust. Math. Soc. 82 (2010) 437–445.

[28] J.A. Cid, D. Franco, F. Minhós, Positive fixed points and fourth-order equations, Bull. Lond. Math. Soc. 41 (1) (2009) 72–78.

[29] J.A. Cid, On extremal fixed points in Schauder's theorem with applications to differential equations, Bull. Belg. Math. Soc. Simon Stevin 11 (1) (2004) 15–20.

[30] W.A. Coppel, Disconjugacy, Lecture Notes in Math., vol. 220, Springer-Verlag, Berlin, New York, 1971.

[31] C. De Coster, P. Habets, An overview of the method of lower and upper solutions for ODEs, in: Nonlinear Analysis and Its Applications to Differential Equations, Lisbon, 1998, in: Progr. Nonlinear Differential Equations Appl., vol. 43, Birkhäuser Boston, Boston, MA, 2001, pp. 3–22.

[32] M. Cherpion, C. De Coster, P. Habets, A constructive monotone iterative method for second-order BVP in the presence of lower and upper solutions, Appl. Math. Comput. 123 (1) (2001) 75–91.

[33] C. De Coster, P. Habets, The lower and upper solutions method for boundary value problems, in: Handbook of Differential Equations, Elsevier/North-Holland, Amsterdam, 2004, pp. 69–160.

[34] C. De Coster, P. Habets, Two-Point Boundary Value Problems: Lower and Upper Solutions, Math. Sci. Eng., vol. 205, Elsevier B. V., Amsterdam, 2006.

[35] C. De Coster, M. Henrard, Existence and localization of solution for second order elliptic BVP in presence of lower and upper solutions without any order, J. Differential Equations 145 (2) (1998) 420–452.

[36] K. Deimling, Nonlinear Functional Analysis, Springer-Verlag, Berlin, 1985.

[37] E. Esmailzadeh, G. Nakhaie-Jazar, Periodic solutions of a Mathieu-Duffing type equation, J. Non.-Linear Mech. 32 (1997) 905–912.

[38] W. Feng, G. Zhang, New fixed point theorems on order intervals and their applications, Fixed Point Theory Appl. 2015 (2015) 218, 10 pp.

[39] D. Franco, G. Infante, J. Perán, A new criterion for the existence of multiple solutions in cones, Proc. Roy. Soc. Edinburgh Sect. A 142 (2012) 1043–1050.

[40] W. Gao, J. Wang, On a nonlinear second order periodic boundary value problem with Carathéodory functions, Ann. Polon. Math. 62 (3) (1995) 283.

[41] J.R. Graef, L. Kong, H. Wang, Existence, multiplicity, and dependence on a parameter for a periodic boundary value problem, J. Differential Equations 245 (2008) 1185–1197.

[42] J.R. Graef, L. Kong, H. Wang, A periodic boundary value problem with vanishing Green's function, Appl. Math. Lett. 21 (2008) 176–180.

[43] J.R. Graef, C. Qian, B. Yang, Multiple symmetric positive solutions of a class of boundary value problems for higher order ordinary differential equations, Proc. Amer. Math. Soc. 131 (2003) 577–585.

[44] A. Granas, J. Dugundji, Fixed Point Theory, Springer Monogr. Math., Springer-Verlag, New York, 2003.

[45] P. Habets, R.L. Pouso, Examples of the nonexistence of a solution in the presence of upper and lower solutions, ANZIAM J. 44 (4) (2003) 591–594.

[46] S. Heikkilä, On fixed points through iteratively generated chains with applications to differential equations, J. Math. Anal. Appl. 138 (2) (1989) 397–417.

[47] S. Heikkilä, Applications of a recursion method to maximization problems in partially ordered spaces, Nonlinear Stud. 3 (2) (1996) 249–260.

[48] S. Heikkilä, V. Lakshmikantham, Monotone Iterative Techniques for Discontinuous Nonlinear Differential Equations, Monogr. Textb. Pure Appl. Math., vol. 181, Marcel Dekker, Inc., New York, 1994.

[49] G. Infante, J.R.L. Webb, Nonzero solutions of Hammerstein integral equations with discontinuous kernels, J. Math. Anal. Appl. 272 (2002) 30–42.

[50] G. Infante, J.R.L. Webb, Three point boundary value problems with solutions that change sign, J. Integral Equations Appl. 15 (2003) 37–57.

[51] G. Infante, J.R.L. Webb, Nonlinear nonlocal boundary value problems and perturbed Hammerstein integral equations, Proc. Edinb. Math. Soc. 49 (2006) 637–656.

[52] G. Infante, J.R.L. Webb, Loss of positivity in a nonlinear scalar heat equation, NoDEA Nonlinear Differential Equations Appl. 13 (2006) 249–261.

[53] G.L. Karakostas, P.Ch. Tsamatos, Existence of multiple positive solutions for a nonlocal boundary value problem, Topol. Methods Nonlinear Anal. 19 (2002) 109–121.

[54] J.L. Kelley, General Topology, Grad. Texts in Math., vol. 27, Springer-Verlag, New York, Berlin, 1975. Reprint of the 1955 edition [Van Nostrand, Toronto, Ont.].

[55] I.T. Kiguradze, A priori estimates for the derivatives of bounded functions satisfying second-order differential inequalities, Differ. Uravn. 3 (1967) 1043–1052 (in Russian).

[56] I.T. Kiguradze, On some singular boundary value problems for the second order nonlinear ordinary differential equations, Differ. Uravn. 4 (10) (1968) 1753–1773 (in Russian).

[57] L. Kong, D. Jiang, Multiple positive solutions of a nonlinear fourth order periodic boundary value problem, Ann. Polon. Math. 69 (1998) 265–270.

[58] M.A. Krasnosel'skiĭ, Fixed points of cone-compressing or cone-extending operators, Sov. Math., Dokl. 1 (1960) 1285–1288.

[59] M.A. Krasnosel'skiĭ, Positive Solutions of Operator Equations, Noordhoff, Groningen, 1964.

[60] G.S. Ladde, V. Lakshmikantham, A.S. Vatsala, Monotone Iterative Techniques for Nonlinear Differential Equations, Pitman, Boston, MA, 1985.

[61] K.Q. Lan, Multiple positive solutions of semilinear differential equations with singularities, J. London Math. Soc. 63 (2001) 690–704.

[62] L.A. Lepin, On the notions of lower and upper functions, Differ. Equ. 16 (10) (1980) 1750–1759 (in Russian).

[63] L.A. Lepin, Generalized lower and upper functions and their properties, Latvian Math. Annu. 24 (1980) 113–123 (in Russian).

[64] L.A. Lepin, Generalized solutions and the solvability of boundary-value problems for a second-order differential equation, Differ. Equ. 18 (8) (1982) 925–931.

[65] A.Ya. Lepin, L.A. Lepin, V.D. Ponomarev, Generalized solvability of nonlinear boundary-value problems, Differ. Equ. 22 (2) (1986) 142–146.

[66] A.Ya. Lepin, F.Ž. Sadyrbaev, The upper and lower functions method for second order systems, Z. Anal. Anwend. 20 (3) (2001) 739–753.

[67] J. Leray, J. Schauder, Topologie et équations fonctionnelles, Ann. Sci. Éc. Norm. Supér. (3) 51 (1934) 45–78 (in French).

[68] N.G. Lloyd, Degree Theory, Cambridge Tracts in Math., vol. 73, Cambridge University Press, Cambridge, New York, Melbourne, 1978.

[69] R. Ma, Multiplicity of positive solutions for second order three-point boundary value problems, Comput. Math. Appl. 40 (2000) 193–204.

[70] J.L. Mawhin, Equivalence theorems for nonlinear operators equations and coincidence degree theory for some mappings in locally convex topological vector spaces, J. Differential Equations 12 (1972) 610–632.

[71] J. Mawhin, Points fixes, points critiques et problèmes aux limites, Sémin. Math. Supér., vol. 92, Presses de l'Université de Montréal, Montreal, QC, 1985, 162 pp. (in French).

[72] J. Mawhin, Twenty years of ordinary differential equations through twelve Oberwolfach meetings, Results Math. 21 (1–2) (1992) 165–189.

[73] J. Mawhin, Boundary value problems for nonlinear ordinary differential equations: from successive approximations to topology, in: Development of Mathematics 1900–1950, Luxembourg, 1992, Birkhäuser, Basel, 1994, pp. 443–477.

[74] J. Mawhin, Bounded solutions of nonlinear ordinary differential equations, in: Non-linear Analysis and Boundary Value Problems for Ordinary Differential Equations (Udine), in: CISM Courses and Lectures, vol. 371, Springer, Vienna, 1996, pp. 121–147.

[75] E.J. McShane, Integration, Princeton University Press, Princeton, 1967.

[76] D.S. Mitrinović, J.E. Pečarić, A.M. Fink, Inequalities Involving Functions and Their Integrals and Derivatives, Math. Appl., East Eur. Ser., vol. 53, Kluwer Academic Publishers Group, Dordrecht, 1991.

[77] M. Müller, Über das Fundamentaltheorem in der theorie der gewöhnlichen differentialgleichungen, Math. Z. 26 (1926) 619–649.

[78] M. Nagumo, Uber die differentialgleichung $y'' = f(t, y, y')$, Proc. Phys.-Math. Soc. Japan 19 (1937) 861–866.

[79] M. Nagumo, On principally linear elliptic differential equations of the second order, Osaka Math. J. 6 (1954) 207–229.

[80] J.J. Nieto, An abstract monotone iterative technique, Nonlinear Anal. 28 (1997) 1923–1933.

[81] P. Omari, A monotone method for constructing extremal solutions of second order scalar boundary value problems, Appl. Math. Comput. 18 (1986) 257–275.

[82] R. Ortega, Degree Theory and Boundary Value Problems, Trieste, 2006, http://www.ugr.es/~rortega/PDFs/degree.pdf.

[83] O. Perron, Ein neuer existenzbeweis für die integrale der differentialgleinchung $y' = f(t, y)$, Math. Ann. 76 (1915) 471–484.

[84] H. Persson, A Fixed Point Theorem for Monotone Functions Equivalent to Brouwer Theorem, Working Paper Series, vol. 1, University of Örebro, 2005. Available at http://www.oru.se/csi/wps.

[85] H. Persson, A fixed point theorem for monotone functions, Appl. Math. Lett. 19 (2006) 1207–1209.

[86] E. Picard, Sur l'application des métodes d'approximations succesives à l'étude de certains équations différentielles ordinaires, J. Math. 9 (1893) 217–271.

[87] E. Picard, Mémoire sur la théorie des équations aux derivés partielles et las méthode des approximations succesives, J. Math. 6 (1890) 145–210.

[88] V.D. Ponomarev, Existence of a solution in a simplest boundary value problem for the second order differential equation, Latvian Math. Annu. 22 (1978) 69–74 (in Russian).

[89] H. Schaefer, Über die Methode der a priori Schranken, Math. Ann. 129 (1955) 415–416.

[90] K.W. Schrader, Solutions of second order ordinary differential equations, J. Differential Equations 4 (1968) 510–518.

[91] J. Schauder, Der Fixpunktsatz in Funktionalriiumen, Studia Math. 2 (1930) 171–180.

[92] S. Scorza Dragoni, Il problema dei valori ai limiti estudiato in grande per le equazione differenziale del secondo ordine, Math. Ann. 105 (1931) 133–143.

[93] S. Scorza Dragoni, Elementi uniti di transformazioni funzionali e problemi di valori ai limiti, Rend. Sem. Mat. Univ. Roma Ser. IV 2 (4) (1938) 255–275.

[94] P.J. Torres, Existence of one-signed periodic solutions of some second-order differential equations via a Krasnosel'skiĭ fixed point theorem, J. Differential Equations 190 (2003) 643–662.

[95] P.J. Torres, Weak singularities may help periodic solutions to exist, J. Differential Equations 232 (2007) 277–284.

[96] H. Wang, On the number of positive solutions of nonlinear systems, J. Math. Anal. Appl. 281 (1) (2003) 287–306.

[97] M.X. Wang, A. Cabada, J.J. Nieto, Monotone method for nonlinear second order periodic boundary value problems with Carathéodory functions, Ann. Polon. Math. 58 (1993) 221–235.

[98] E. Zeidler, Nonlinear Functional Analysis and Its Applications. I. Fixed-Point Theorems, Springer-Verlag, New York, 1986.

[99] M. Zhang, A relationship between the periodic and the Dirichlet BVPs of singular differential equations, Proc. Roy. Soc. Edinburgh Sect. A 128 (1998) 1099–1114.

[100] G. Zhang, J. Sun, A generalization of the cone expansion and compression fixed point theorem and applications, Nonlinear Anal. 67 (2007) 579–586.

APPENDIX A

Sobolev Inequalities

Throughout this book, we have used several Sobolev inequalities. For the sake of completeness we will include now their proof.

It is well-known that for p, $q > 1$, the embedding $W_0^{1,p}(I)$ into $L^q(I)$ is compact (see [2, Chapter 9]). In particular, for any $u \in W_0^{1,p}(I)$, the following inequality holds

$$\|u\|_{L^q(I)} \le c_{pq} \|u'\|_{L^p(I)} \tag{A.1}$$

with

$$c_{pq} = \frac{T^{\frac{1}{q}+\frac{1}{p^*}}}{2B\left(\frac{1}{q},\frac{1}{p^*}\right)} \, (p^*)^{\frac{1}{q}} \, (q)^{\frac{1}{p^*}} \, (p^*+q)^{\frac{1}{p}-\frac{1}{q}}, \tag{A.2}$$

where (A.2) is the best possible constant in (A.1), B denotes the classical Beta function and p^* denotes the conjugate of p, that is, the number satisfying the relation $\frac{1}{p} + \frac{1}{p^*} = 1$ (with $p = 1$ and $p^* = \infty$ and vice-versa).

We will include now a proof of this result. We will follow the arguments made by P. Drábek and R. Manásevich in [6] for the case $1 < p < \infty$ and the calculations made by A. Cabada, J.A. Cid and M. Tvrdý in [4] for the cases $q = 1$ and $q = \infty$.

We note that this expression was first given by G. Talenti in 1976, [11]. He proved the result in $\mathbb{R}^n$ for $n > 1$ and he indicated that the constant c_{pq} is optimal in the case $n = 1$, despite he did not prove it.

It is clear that

$$\frac{1}{c_{pq}} = \inf_{u \in W_0^{1,p}(I)\setminus\{0\}} \frac{\left(\int_0^T |u'(t)|^p \, dt\right)^{\frac{1}{p}}}{\left(\int_0^T |u(t)|^q \, dt\right)^{\frac{1}{q}}}.$$

To solve this equation we will consider a two steps variational method:

1. First we will solve the isoperimetric problem of minimizing the functional

$$J[u] = \int_0^T |u'(t)|^p \, dt \tag{A.3}$$

Maximum Principles for the Hill's Equation.
DOI: http://dx.doi.org/10.1016/B978-0-12-804117-8.00013-8

subject to the boundary conditions

$$u(0) = u(T) = 0 \tag{A.4}$$

and the restriction that

$$K[u] \equiv \int_0^T |u(t)|^q \, dt = c^q. \tag{A.5}$$

2. If, for every positive constant c, the previous problem has a minimum $m(c)$, then it is clear that

$$\frac{1}{c_{pq}} = \inf_{c>0} \frac{(m(c))^{\frac{1}{p}}}{c}. \tag{A.6}$$

Let's consider then the problem of minimizing (A.3) under conditions (A.4) and (A.5). To solve this problem we use the Lagrange's multiplier technique (see for instance [10]). Therefore we know that the minimum u we are looking for is such that the function

$$f(t, u, u') \equiv |u'|^p - \mu \, |u|^q$$

satisfies the Euler-Lagrange's equation, that is,

$$\frac{\partial f(t, u, u')}{\partial u} - \frac{d}{dt} \left[\frac{\partial f(t, u, u')}{\partial u'} \right] \equiv -\mu \, q |u|^{q-2} u - p \left(|u'|^{p-2} u' \right)' = 0$$

or, which is the same, u needs to be a solution of the equation

$$(\phi_p(u'(t)))' + \frac{\mu \, q}{p} \phi_q(u(t)) = 0, \quad \text{a.e. } t \in I, \tag{A.7}$$

with

$$\phi_m(s) = \begin{cases} |s|^{m-2} s, & s \in \mathbb{R} \setminus \{0\}, \\ 0, & s = 0, \end{cases}$$

$m > 1$.

Moreover, u must satisfy (A.4) and (A.5).

Equation

$$(\phi_p(u'(t)))' + \lambda \, \phi_q(u(t)) = 0, \quad \text{a.e. } t \in I,$$

where λ is a positive parameter, was studied by P. Drábek and R. Maná-sevich in [6]. We will follow their results. First they considered the initial value problem

$$\begin{cases} (\phi_p(u'(t)))' + \lambda\,\phi_q(u(t)) = 0, \text{ a.e. } t \in \mathbb{R}, \\ u(t_0) = a, \ u'(t_0) = b, \ t_0 \in \mathbb{R} \end{cases} \tag{A.8}$$

and proved the following existence result.

Proposition 5. *[6, Proposition 2.1] For any $\lambda \in [0, \infty)$, the initial value problem (A.8) has a unique solution defined in $\mathbb{R}$.*

Then, they considered the particular case of

$$\begin{cases} (\phi_p(u'(t)))' + \lambda\,\phi_q(u(t)) = 0, \text{ a.e. } t \in \mathbb{R}, \\ u(0) = 0, \ u'(0) = \alpha, \end{cases} \tag{A.9}$$

with α a positive constant.

Now, if u is a solution of (A.9) and $t(\alpha)$ denotes the first zero of u', it is clear that both u and u' are positive on $(0, t(\alpha))$. Therefore multiplying both sides of the differential equation in (A.9) by u' and integrating on $[0, t]$ for $t \in [0, t(\alpha)]$ we obtain that

$$\frac{(u'(t))^p}{p^*} + \lambda\,\frac{(u(t))^q}{q} = \frac{\alpha^p}{p^*}, \quad t \in [0, t(\alpha)]. \tag{A.10}$$

The same way, integrating now the same expression on $[t, t(\alpha)]$, we obtain

$$\frac{(u'(t))^p}{p^*} + \lambda\,\frac{(u(t))^q}{q} = \lambda\,\frac{R^q}{q}, \quad t \in [0, t(\alpha)], \tag{A.11}$$

with $R = u(t(\alpha))$.

Previous equation is equivalent to

$$\left(\frac{q}{\lambda\,p^*}\right)^{\frac{1}{p}} \frac{u'(t)}{(R^q - (u(t))^q)^{\frac{1}{p}}} = 1$$

and integrating on $(0, t)$ we get

$$\left(\frac{q}{\lambda\,p^*}\right)^{\frac{1}{p}} \int_0^t \frac{u'(s)}{(R^q - (u(s))^q)^{\frac{1}{p}}}\,ds = t, \quad t \in [0, t(\alpha)],$$

or, which is the same,

$$t = \left(\frac{q}{\lambda\,p^*}\right)^{\frac{1}{p}} \frac{1}{R^{\frac{q-p}{p}}} \int_0^{\frac{u(t)}{R}} \frac{1}{(1 - s^q)^{\frac{1}{p}}}\,ds, \quad t \in [0, t(\alpha)]. \tag{A.12}$$

Now, for $\sigma \in \left[0, \frac{q}{2}\right]$, we define

$$\operatorname{arcsin}_{pq}(\sigma) := \frac{q}{2} \int_0^{\frac{2\sigma}{q}} \frac{1}{(1 - s^q)^{\frac{1}{p}}} \, ds = \frac{1}{2} \int_0^{\left(\frac{2\sigma}{q}\right)^q} z^{\frac{1}{q} - 1} (1 - z)^{-\frac{1}{p}} \, dz.$$

In particular, we will denote

$$\pi_{pq} := 2 \operatorname{arcsin}_{pq}\left(\frac{q}{2}\right) = B\left(\frac{1}{q}, \frac{1}{p^*}\right),$$

with B the Beta function.

We note that $\operatorname{arcsin}_{pq}(\sigma)$ is well defined for $\sigma \in \left[0, \frac{q}{2}\right]$ and satisfies that

$$\operatorname{arcsin}_{pq} : \left[0, \frac{q}{2}\right] \to \left[0, \frac{\pi_{pq}}{2}\right]$$

is strictly increasing. Therefore, it is invertible and its inverse, which will be denoted by $\sin_{pq}$, is also strictly increasing on $\left[0, \frac{\pi_{pq}}{2}\right]$.

This function $\sin_{pq} : \left[0, \frac{\pi_{pq}}{2}\right] \to \left[0, \frac{q}{2}\right]$ can be extended to the whole real line (and we will denote its extension also as $\sin_{pq}$) in the following way:

1. First, for $t \in \left[\frac{\pi_{pq}}{2}, \pi_{pq}\right]$, we define $\sin_{pq}(t) = \sin_{pq}(\pi_{pq} - t)$.
2. Then, for $t \in [-\pi_{pq}, 0]$, we define $\sin_{pq}(t) = -\sin_{pq}(-t)$.
3. Finally, we consider the $2\pi_{pq}$ extension of the function to the whole real line.

Now, taking into account the definition of $\sin_{pq}$, we rewrite (A.12) as

$$t = \frac{2}{(\lambda p^*)^{\frac{1}{p}} q^{\frac{1}{p^*}}} R^{\frac{p-q}{p}} \operatorname{arcsin}_{pq}\left(\frac{q\, u(t)}{2R}\right)$$

and hence

$$u(t) = \frac{2R}{q} \sin_{pq}\left(\frac{(\lambda p^*)^{\frac{1}{p}} q^{\frac{1}{p^*}}}{2} R^{\frac{q-p}{p}} t\right)$$

for all $t \in \left[0, \frac{\pi_{pq}}{2}\right]$.

Due to the extension of $\sin_{pq}$ that we have just made, previous expression is valid for all $t \in \mathbb{R}$.

On the other hand, from (A.10) and (A.11), we obtain that

$$R = \left(\frac{q}{\lambda p^*}\right)^{\frac{1}{q}} \alpha^{\frac{p}{q}}.$$

Therefore, we conclude that the general solution of problem (A.9) is

$$u(t) = \frac{2\alpha^{\frac{p}{q}}}{(\lambda p^*)^{\frac{1}{q}} q^{\frac{1}{q^*}}} \sin_{pq}\left(\frac{(\lambda p^*)^{\frac{1}{q}} q^{\frac{1}{q^*}}}{2} \alpha^{\frac{q-p}{q}} t\right), \quad t \in \mathbb{R}. \tag{A.13}$$

Now, we have the following result

Theorem 75. *[6, Theorem 3.1] For any given $\alpha \neq 0$, the set of eigenvalues of problem*

$$\begin{cases} (\phi_p(u'(t)))' + \lambda\,\phi_q(u(t)) = 0, \; a.e. \; t \in I, \\ u(0) = 0, \; u(T) = 0, \end{cases} \tag{A.14}$$

is given by

$$\lambda_n(\alpha) = \left(\frac{2\,n\,\pi_{pq}}{T}\right)^q \frac{|\alpha|^{p-q}}{p^*\,q^{q-1}}, \quad n = 1, 2, 3, \ldots,$$

with corresponding eigenfunctions

$$u_{n,\alpha}(t) = \frac{\alpha\,T}{n\,\pi_{pq}} \sin_{pq}\left(\frac{n\,\pi_{pq}}{T} t\right).$$

Proof. The result follows immediately from imposing condition $u(T) = 0$ to the general solution of (A.9) given in (A.13). $\qquad\square$

Therefore, the eigenvalues of Eq. (A.7) coupled with Dirichlet conditions $u(0) = u(T) = 0$ can be obtained just by considering the ones for problem (A.14) multiplied by $\frac{p}{q}$, that is

Corollary 48. *For any given $\alpha \neq 0$, the set of eigenvalues of problem*

$$\begin{cases} (\phi_p(u'(t)))' + \dfrac{\mu\,q}{p}\,\phi_q(u(t)) = 0, \; a.e. \; t \in I, \\ u(0) = 0, \; u(T) = 0, \end{cases}$$

is given by

$$\mu_n(\alpha) = (p-1)\left(\frac{2\,n\,\pi_{pq}}{T}\right)^q \frac{|\alpha|^{p-q}}{q^q}, \quad \text{for each } n \in \mathbb{N},$$

with corresponding eigenfunctions

$$u_{n,\alpha}(t) = \frac{\alpha\,T}{n\,\pi_{pq}} \sin_{pq}\left(\frac{n\,\pi_{pq}}{T} t\right).$$

Consider now $w_n(t) = \frac{1}{n}\sin_{pq}(n\,t)$, $n \in \mathbb{N}$, which is the solution to the following boundary value problem

$$\begin{cases} (\phi_p(w_n'(t)))' + \dfrac{(2\,n)^q}{p^*\,q^{q-1}}\,\phi_q(w_n(t)) = 0, & \text{a.e. } t \in (0, \pi_{pq}), \\ w_n(0) = 0, \quad w_n(\pi_{pq}) = 0. \end{cases}$$

We observe that these solutions satisfy that $w_n'(0) = 1$.

Multiplying each side of the previous differential equation by w_n and integrating on $[0, \pi_{pq}]$, we get

$$\int_0^{\pi_{pq}} |w_n'(t)|^p\, dt = \frac{(2\,n)^q}{p^*\,q^{q-1}} \int_0^{\pi_{pq}} |w_n(t)|^q\, dt. \tag{A.15}$$

On the other hand, from (A.10), since w_n satisfies that

$$\frac{|w_n'(t)|^p}{p^*} + \frac{(2\,n)^q}{p^*\,q^{q-1}}\,\frac{|w_n(t)|^q}{q} = \frac{1}{p^*},$$

we integrate this expression on $[0, \pi_{pq}]$ and we obtain

$$\frac{1}{p^*}\int_0^{\pi_{pq}} |w_n'(t)|^p\, dt + \frac{(2\,n)^q}{p^*\,q^q} \int_0^{\pi_{pq}} |w_n(t)|^q\, dt = \frac{\pi_{pq}}{p^*}. \tag{A.16}$$

Consequently, from (A.15) and (A.16) we deduce that

$$\int_0^{\pi_{pq}} |w_n(t)|^q\, dt = \frac{\pi_{pq}\,p^*\,q^q}{(2\,n)^q\,(p^* + q)}$$

and

$$\int_0^{\pi_{pq}} |w_n'(t)|^p\, dt = \frac{\pi_{pq}\,q}{(p^* + q)}.$$

From the previous expressions, making a suitable change of variable, it follows that

$$\int_0^T \left|\sin_{pq}\left(\frac{n\,\pi_{pq}}{T}\,t\right)\right|^q\, dt = \frac{T\,n^q}{\pi_{pq}} \int_0^{\pi_{pq}} |w_n(t)|^q\, dt = \frac{T\,p^*\,q^q}{2^q\,(p^* + q)}$$

and

$$\int_0^T \left|\frac{d}{dt}\sin_{pq}\left(\frac{n\,\pi_{pq}}{T}\,t\right)\right|^p\, dt = n^p\,\frac{\pi_{pq}^{p-1}}{T^{p-1}} \int_0^{\pi_{pq}} |w_n'(t)|^p\, dt = \frac{n^p\,\pi_{pq}^p\,q}{T^{p-1}\,(p^* + q)}.$$

Finally we need to impose condition (A.5) to $u_{n,\alpha}$, that is,

$$c^q = K[u_{n,\alpha}(t)] = \int_0^T |u_{n,\alpha}(t)|^q \, dt = \left(\frac{\alpha \, T}{n \, \pi_{pq}}\right)^q \int_0^T \left|\sin_{pq}\left(\frac{n \, \pi_{pq}}{T} t\right)\right|^q \, dt$$

$$= \left(\frac{\alpha \, T}{n \, \pi_{pq}}\right)^q \frac{T \, p^* \, q^q}{2^q \, (p^* + q)},$$

from where we obtain that

$$\alpha = c \frac{2}{q} \left(\frac{p^* + q}{T \, p^*}\right)^{\frac{1}{q}} \frac{n \, \pi_{pq}}{T}$$

and therefore

$$u_{n,\alpha}(t) = c \frac{2}{q} \left(\frac{p^* + q}{T \, p^*}\right)^{\frac{1}{q}} \sin_{pq}\left(\frac{n \, \pi_{pq}}{T} t\right).$$

As a consequence, taking into account (A.6), we have that

$$(m(c))^{\frac{1}{p}} = c \frac{2}{q} \left(\frac{p^* + q}{T \, p^*}\right)^{\frac{1}{q}} \left(\int_0^T \left|\frac{d}{dt} \sin_{pq}\left(\frac{n \, \pi_{pq}}{T} t\right)\right|^p \, dt\right)^{\frac{1}{p}}$$

$$= c \frac{2}{q} \left(\frac{p^* + q}{T \, p^*}\right)^{\frac{1}{q}} \left(\frac{n^p \, \pi_{pq}^p \, q}{T^{p-1} (p^* + q)}\right)^{\frac{1}{p}}.$$

In particular we note that $\frac{(m(c))^{\frac{1}{p}}}{c}$ does not depend on c and we deduce that

$$\frac{1}{c_{pq}} = \min_{n=1,2,\dots} \frac{2}{q} \left(\frac{p^* + q}{T \, p^*}\right)^{\frac{1}{q}} \left(\frac{n^p \, \pi_{pq}^p \, q}{T^{p-1} (p^* + q)}\right)^{\frac{1}{p}}$$

$$= \frac{2 \, \pi_{pq}}{q \, T^{\frac{1}{q} + \frac{1}{p^*}}} \left(\frac{p^* + q}{p^*}\right)^{\frac{1}{q}} \left(\frac{q}{p^* + q}\right)^{\frac{1}{p}},$$

or which is the same,

$$c_{pq} = \frac{T^{\frac{1}{q} + \frac{1}{p^*}}}{2 \, \pi_{pq}} \, (p^*)^{\frac{1}{q}} \, (q)^{\frac{1}{p^*}} \, (p^* + q)^{\frac{1}{p} - \frac{1}{q}},$$

as we wanted to prove.

Remark 51. We note that, following the steps of [6] and [11], we have defined the Sobolev constant as given in Eq. (A.1). However, in the recent literature (and in particular in papers related with Hill's equation) it is common to define $K(q, p, T)$ as the best Sobolev constant in the inequality

$$C \|u\|_{L^q(I)}^p \leq \|u'\|_{L^p(I)}^p \quad \text{for all } u \in W_0^{1,p}(I),$$

that is

$$K(q, p, T) = \inf \left\{ \frac{\|u'\|_{L^p(I)}^p}{\|u\|_{L^q(I)}^p} : u \in W_0^{1,p}(I) \setminus \{0\} \right\}.$$

With our notation, it is clear that

$$K(q, p, T) = \left(\frac{1}{c_{pq}} \right)^p.$$

Finally, we will see how A. Cabada, J.A. Cid and M. Tvrdý calculated in [4] the value of $K(q, p, T)$ for $q = 1$ and $q = \infty$.

Put

$$\varkappa(q, p) = \left(\frac{2 \left(1 + \frac{q}{p^*}\right)^{1/q} B\left(\frac{1}{q}, \frac{1}{p^*}\right)}{q \left(1 + \frac{p^*}{q}\right)^{1/p}} \right)^p \quad \text{for } q \in [1, \infty) \quad \text{and} \quad p \in (1, \infty).$$

We have just proved that

$$K(q, p, T) = \frac{\varkappa(q, p)}{T^{p-1+p/q}} \quad \text{for } 1 < q < \infty.$$

Furthermore, one can show that the relations

$$K(1, p, T) = \frac{\varkappa(1, p)}{T^{2p-1}} \quad \text{and} \quad K(\infty, p, T) = \frac{2^p}{T^{p-1}}$$

are true, as well.

Case $q = 1$. Letting $q \to 1$ in the relation

$$K(q, p, T) \|u\|_{L^q(I)}^p \leq \|u'\|_{L^p(I)}^p \quad \text{for all } u \in W_0^{1,p}(I), \tag{A.17}$$

we obtain

$$\frac{\varkappa(1, p)}{T^{2p-1}} \|u\|_{L^1(I)}^p \leq \|u'\|_{L^p(I)}^p \quad \text{for all } u \in W_0^{1,p}(I). \tag{A.18}$$

Moreover, as it can be seen in [13, Remark 2.2 (i)], the equality in (A.18) is achieved by some $u \in W_0^{1,p}(I)$. This proves the equality $K(1, p, T) = \dfrac{\varkappa(1, p)}{T^{2p-1}}$.

Case $q = \infty$.   First, recall that

$$\lim_{q \to \infty} \|u\|_{L^q(I)}^p = \|u\|_{L^\infty(I)}^p$$

holds for all $u \in W_0^{1,p}(I)$ and all $p \in (1, \infty)$ (cf. e.g. [12, Theorem I.3.1]). Therefore, letting $q \to \infty$ in (A.17), we get

$$\frac{2^p}{T^{p-1}} \|u\|_{L^\infty(I)}^p \le \|u'\|_{L^p(I)}^p \quad \text{for all } u \in W_0^{1,p}(I).$$

On the other hand, let $C > 0$ be an arbitrary constant such that

$$C \|u\|_{L^\infty(I)}^p \le \|u'\|_{L^p(I)}^p \quad \text{for all } u \in W_0^{1,p}(I).$$

Since $\|u\|_{L^q(I)}^p \le \|u\|_{L^\infty(I)}^p T^{p/q}$, we have

$$\frac{C}{T^{p/q}} \|u\|_{L^q(I)}^p \le C \|u\|_{L^\infty(I)}^p \le \|u'\|_{L^p(I)}^p \quad \text{for all } u \in W_0^{1,p}(I),$$

and, by the definition of $K(q, p, T)$, it follows that

$$\frac{C}{T^{p/q}} \le K(q, p, T).$$

Thus, letting $q \to \infty$, we get

$$C \le \frac{2^p}{T^{p-1}},$$

wherefrom the equality $K(\infty, p, T) = \dfrac{2^p}{T^{p-1}}$ immediately follows.
    Therefore we conclude that

$$K(q, p, T) = \begin{cases} \dfrac{\left(2\left(1+\frac{q}{p^*}\right)^{1/q} B\left(\frac{1}{q}, \frac{1}{p^*}\right)\right)^p}{q^p\, T^{p-1+p/q}\left(1+\frac{p^*}{q}\right)}, & \text{if } 1 \le q < \infty, \\[4mm] \dfrac{2^p}{T^{p-1}}, & \text{if } q = \infty. \end{cases}$$

Remark 52. As it has been pointed out in [5], we note that the main step to find the general solution of problem (A.9) (and consequently the eigenfunctions of problem (A.14)) is the definition of the function $\sin_{pq}$. This notation comes from the fact that if we consider the equation

$$(g \circ u')'(t) + f(u(t)) = 0, \qquad u(0) = 0, \quad u'(0) = 1, \tag{A.19}$$

it is clear that, in the case $g(u) = f(u) = u$, the unique solution of problem (A.19) is $\sin(t)$. This suggests the definition of the $\sin_{g,f}$ function as the unique solution of problem (A.19) for general g and f.

This function, defined as such, coincides with the $\sin_p$ function defined in [3,9] for the p-Laplacian $f(u) = g(u) = |u|^{p-2}u$, the function $\sin_{pq}$ defined in [1,7,8] for the p-q-Laplacian $f(u) = |u|^{q-2}u$, $g(u) = |u|^{p-2}u$, and the hyperbolic version of this function, also in [1,8], which corresponds to the case $f(u) = |u|^{q-2}u$, $g(u) = -|u|^{p-2}u$. The definition that we have used is the one considered in [6], which is previous and slightly different to the one given in [1,7,8] but still follows the same idea.

On the other hand, A. Cabada and F.A.F. Tojo defined in [5] the function $\sin_{g,f}$ with f and g not necessarily odd functions, contrary to the above examples. Such definition covers all the previous situations cited before.

REFERENCES

[1] B.A. Bhayo, M. Vuorinen, On generalized trigonometric functions with two parameters, J. Approx. Theory 164 (10) (2012) 1415–1426.

[2] H. Brezis, Functional Analysis, Sobolev Spaces and Partial Differential Equations, Springer, New York, 2010.

[3] S.N. Busenberg, C.C. Travis, On the use of reducible-functional differential equations in biological models, J. Math. Anal. Appl. 89 (1) (1982) 46–66.

[4] A. Cabada, J.A. Cid, M. Tvrdý, A generalized anti-maximum principle for the periodic one-dimensional p-Laplacian with sign-changing potential, Nonlinear Anal. 72 (7–8) (2010) 3436–3446.

[5] A. Cabada, F.A.F. Tojo, Periodic solutions for some phi-Laplacian and reflection equations, Bound. Value Probl. 2016 (2016) 56, 16 pp.

[6] P. Drábek, R. Manásevich, On the closed solution to some nonhomogeneous eigenvalue problems with p-Laplacian, Diff. Int. Eq. 12 (6) (1999) 773–788.

[7] D.E. Edmunds, P. Gurka, J. Lang, Properties of generalized trigonometric functions, J. Approx. Theory 164 (1) (2012) 47–56.

[8] W.D. Jiang, M.K. Wang, Y.M. Chu, Y.P. Jiang, F. Qi, Convexity of the generalized sine function and the generalized hyperbolic sine function, J. Approx. Theory 174 (2013) 1–9.

[9] R. Klén, M. Vuorinen, X. Zhang, Inequalities for the generalized trigonometric and hyperbolic functions, J. Math. Anal. Appl. 409 (1) (2014) 521–529.

[10] M. Kot, A First Course in the Calculus of Variations, American Mathematical Society, 2014.

[11] G. Talenti, Best constant in Sobolev inequality, Ann. Mat. Pura Appl. 4 (1976) 353–372.

[12] K. Yosida, Functional Analysis, Springer Verlag, Berlin, Heidelberg, New York, 1978.

[13] M. Zhang, Certain classes of potentials for p-Laplacian to be non-degenerate, Math. Nachr. 278 (15) (2005) 1823–1836.

GLOSSARY

$\mathbb{N}$ Set of natural numbers, that is, $\{1, 2, \dots\}$.

$\mathbb{Z}$ Set of integer numbers.

$\mathbb{R}$ Set of real numbers.

$\mathbb{C}$ Set of complex numbers.

$I = [0, T] = \{t \in \mathbb{R}, \ 0 \leq t \leq T\}$.

$J = [0, 2T] = \{t \in \mathbb{R}, \ 0 \leq t \leq 2T\}$.

$\mathcal{C}(I)$ Space of continuous real functions defined on I.

$\mathcal{C}^k(I)$, $k \in \mathbb{N}$ Space of k-times differentiable real functions defined on I such that the j-th derivative is continuous for $j = 0, \dots, k$.

$\mathcal{C}^\infty(I)$ Space of infinitely differentiable real functions defined on I.

$L^\alpha(I)$, $1 \leq \alpha < \infty$ Space of the measurable functions f on the interval I such that the Lebesgue integral of $|f|^\alpha$ is finite.

$\|f\|_\alpha$, $1 \leq \alpha < \infty$ Norm of f in the space $L^\alpha(I)$, that is,

$$\|f\|_\alpha = \left(\int_I |f(t)|^\alpha \, dt \right)^{\frac{1}{\alpha}}.$$

$L^\infty(I)$ Space of the measurable functions f on the interval I such that are essentially bounded.

$\|f\|_\infty$ Norm of f in the space $L^\infty(I)$, that is,

$$\|f\|_\infty = \sup \text{ ess} \left\{ |f(t)|, \ t \in I \right\}.$$

$AC(I)$ Absolutely continuous functions, that is,

$$AC(I) = \left\{ u \in \mathcal{C}(I) \mid \exists f \in L^1(I), \ u(t) = u(t_0) + \int_{t_0}^t f(s) \, ds, \ t, \ t_0 \in I \right\}.$$

$W^{k,p}(I)$, $k, p \in \mathbb{N}$ Sobolev space $k - p$ on the set I, that is,

$$W^{k,p}(I) = \left\{ u \in \mathcal{C}^{k-1}(I) \mid u^{(k-1)} \in AC(I), \ u^{(k)} \in L^p(I) \right\}.$$

$H_0^1(I)$ Hilbert space of the $W^{1,2}(I)$ functions that satisfy the Dirichlet conditions, that is,

$$H_0^1(I) = \left\{ u \in W^{1,2}(I) \mid u(0) = u(T) = 0 \right\}.$$

$W(y_1, y_2)$ Wronskian of the functions y_1 and y_2, that is,

$$W(y_1, y_2) = y_1 \, y_2' - y_2 \, y_1'.$$

$tr(M)$ Trace of the square matrix M, that is, the sum of the elements of its diagonal.

$u_1(\cdot, \lambda)$ Solution of the Hill's equation with potential $a + \lambda$ such that $u_1(0, \lambda) = 1$ and $u_1'(0, \lambda) = 0$.

$u_2(\cdot, \lambda)$ Solution of the Hill's equation with potential $a + \lambda$ such that $u_2(0, \lambda) = 0$ and $u_2'(0, \lambda) = 1$.

$\Delta(\lambda)$ Discriminant of the Hill's equation with potential $a + \lambda$, that is,

$$\Delta(\lambda) = u_1(T, \lambda) + u_2'(T, \lambda).$$

L[a] Second order operator related to Hill's equation with potential a, that is,

$$L[a]u(t) = u''(t) + a(t)u(t).$$

$\lambda_k^\sharp[a]$, $\sharp \in \{D, P, A, N, M_1, M_2\}$, $k \in \{0\} \cup \mathbb{N}$ $(k+1)$-th eigenvalue of the Hill's equation with potential a and Dirichlet, periodic, antiperiodic, Neumann or mixed boundary conditions, respectively.

$G_\sharp[a]$, $\sharp \in \{D, P, A, N, M_1, M_2\}$ Green's function of the Hill's equation with potential a and Dirichlet, periodic, antiperiodic, Neumann or mixed boundary conditions, respectively.

$X_\sharp$, $\sharp \in \{P, N, M_1, M_2\}$ Subset of functions in $W^{2,1}(I)$ which satisfy periodic, Neumann or mixed boundary conditions, respectively.

$\Lambda_\sharp[a, T]$, $\sharp \in \{D, P, A, N, M_1, M_2\}$ Spectrum of the Hill's equation with potential a and Dirichlet, periodic, antiperiodic, Neumann or mixed boundary conditions, respectively, on the interval $[0, T]$.

$\tilde{a}$ Even extension of $a: [0, T] \to \mathbb{R}$ to the interval $[0, 2T]$.

$\tilde{\tilde{a}}$ Even extension of $\tilde{a}: [0, 2T] \to \mathbb{R}$ to the interval $[0, 4T]$.

a_+ Positive part of $a: [0, T] \to \mathbb{R}$, that is, $a_+(t) = \max\{a(t), 0\}$.

a_- Negative part of $a: [0, T] \to \mathbb{R}$, that is, $a_-(t) = \max\{-a(t), 0\}$.

$\hat{a}$ Mean value zero of a, that is,

$$\hat{a}(t) = a(t) - \frac{1}{T} \int_0^T a(s)\, ds.$$

α^* Conjugate of the real number $\alpha \geq 1$, that is, the real number such that $\frac{1}{\alpha} + \frac{1}{\alpha*} = 1$. If $\alpha = 1$ then $\alpha^* = \infty$ and vice-versa.

$a \succ 0$ $a \in L^\alpha(I)$ such that $a(t) \geq 0$ for a. e. $t \in I$ and $a \not\equiv 0$ on I.

$K(\alpha, T)$ Best Sobolev constant in the inequality $C \|u\|_\alpha^2 \leq \|u'\|_2^2$ for $u \in H_0^1(I)$, given explicitly by

$$K(\alpha, T) = \begin{cases} \dfrac{2\pi}{\alpha\, T^{1+2/\alpha}} \left(\dfrac{2}{2+\alpha}\right)^{1-2/\alpha} \left(\dfrac{\Gamma(1/\alpha)}{\Gamma(1/2 + 1/\alpha)}\right)^2, & \text{if } 1 \leq \alpha < \infty, \\[2ex] \dfrac{4}{T}, & \text{if } \alpha = \infty. \end{cases}$$

$\overline{D}$ Closure of the set D.

∂D Boundary of the set D.

$int(D)$ Interior of the set D.

$d(\phi, D, p)$ Leray-Schauder degree of operator ϕ at p relative to D.

$i(\mathcal{T}, V)$ Index of operator $\mathcal{T}$ relative to V, defined as

$$i(\mathcal{T}, V) = d(I - \mathcal{T}, V, 0).$$

$\mathcal{K}_{u_0}$, with $u_0 \in K$, $\|u_0\| \leq 1$ Subcone of K, defined as

$$\mathcal{K}_{u_0} = \{u \in K : u \geq \|u\| u_0\}.$$

INDEX

A

Amann's three fixed points Theorem, 138
Anti-maximum principle, 49, 50, 107, 129, 214, 217
Antiperiodic boundary conditions, 20, 32, 33, 41–44, 46, 87

C

Carathéodory function, 130, 140, 143, 144, 187–189, 191, 208, 212
Completely continuous operator, 132–134, 136–138, 141, 145, 149–152, 154–156, 159–163, 171, 179, 195, 201, 203
Cone, 129, 135–141, 143, 149, 152, 154–156, 159–163, 165, 168–170, 178, 201

D

Degree, 131, 132, 134, 135, 150, 152, 169
Dini derivatives, 197, 198
Dirichlet boundary conditions, 19, 20, 26, 27, 29, 39, 42–45, 64, 67, 84, 85, 93, 98, 125, 139, 144, 168, 178, 185–187, 192, 194–198, 201–203, 205, 210, 212–214
Downward directed, 148–150

E

Eigenvalue, 20, 29–31, 42–46, 144, 192, 212, 213, 215, 216
Extremal fixed points, 148–151, 155, 156, 163, 210
Extremal solutions, 150, 198–200, 209, 213, 214

F

Finite intersection property, 149, 150, 199, 200
Floquet's theorem, 8, 10

G

Greatest fixed point, 149–151, 201

H

Green's function, 52–63, 66, 67, 79–82, 84, 87, 90, 91, 93, 96, 97, 99, 100, 103, 105, 106, 110, 117–121, 123–126, 129, 130, 139–141, 143–145, 158, 168–170, 177, 178, 184, 192, 195, 201–203, 205, 210, 213, 214, 216–218

H

Hilbert-Schmidt operator, 210, 211, 214

I

Index, 133, 138, 164
Inverse negative operator, 210
Inverse positive operator, 214

K

Krasnosel'skiĭ's fixed point Theorem, 129, 137, 140–142, 168, 171, 179

L

Lattice, 148–150
Lower solution, 129, 139, 151, 155, 158–160, 163, 184–188, 196–203, 205, 208–211, 213–215, 217
Lyapunov's stability criterium, 17

M

Maximal solution, 199–201, 212, 214, 215
Maximum principle, 49, 50, 97, 211, 217
Minimal solution, 200, 201, 211, 212, 214, 215
Mixed boundary conditions, 20, 29, 30, 90, 113, 125, 139, 144, 148, 168, 178, 186, 212–214, 216
Monodromy matrix, 6, 7, 9
Monotone iterative technique, 129, 151, 153–155, 185, 210, 213, 214, 216, 217

N

Nagumo's condition, 186–189, 193, 197

Neumann boundary conditions, 20, 29, 30, 42–44, 81–83, 85, 91, 96, 97, 124, 143, 158, 168, 186, 196, 212–214, 216–218

Non-decreasing operator, 150–157, 163, 166, 207, 210, 211, 214

Non-increasing operator, 159–162, 166, 210

Normal cone, 135, 140, 152–155, 163, 165

P

Periodic boundary conditions, 20, 32, 33, 41–44, 46, 58, 62, 80, 81, 85, 89, 96, 103, 104, 106, 113, 119, 120, 129, 130, 139, 140, 142, 150, 158, 164, 165, 168, 175, 184, 186, 187, 189, 191, 197, 199, 201, 202, 208, 212, 214, 215

S

Schaefer's fixed point Theorem, 134, 195, 196

Schauder's fixed point Theorem, 134, 135, 149, 175, 184, 203

Smallest fixed point, 149, 150

Solid cone, 135, 152, 154, 155, 159, 160, 163, 165

Stability, 4–6, 8, 14, 15, 17–19

Sturm-Liouville boundary conditions, 45, 197

Sturm's comparison theorem, 25, 29, 30, 44, 45

Sturm's separation theorem, 20, 43, 45

U

Upper solution, 129, 139, 151, 155, 158–160, 163, 184–189, 192, 196–203, 205, 208–211, 213–215, 217

Upward directed, 148–150, 201

ﻭ

Printed in the United States
By Bookmasters